THE END
OF THE RIVER

DAMS, DROUGHT AND DÉJÀ VU
ON THE RIO SÃO FRANCISCO

BRIAN HARVEY

ECW PRESS

Published by ECW Press, 2120 Queen Street East, Suite 200
Toronto, Ontario, Canada M4E 1E2
416.694.3348 / info@ecwpress.com

LIBRARY AND ARCHIVES CANADA CATALOGUING IN PUBLICATION

Harvey, Brian J., 1948 –
  The end of the river : dams, drought and déjà vu on
the Rio São Francisco / Brian Harvey.

ISBN 978-1-55022-845-8

  1. Stream ecology. 2. Stream conservation. 3. Fisheries —
Environmental aspects. I. Title.

QH545.F53H37 2008   333.91'62   C2008-902416-8

Cover and Text Design: Tania Craan
Typesetting: Mary Bowness
Printing: Webcom

Photo Credits:
All photographs are by Brian Harvey except: Rio Negro family (David Greer); sockeye salmon (Carmen Ross); Iguaçu Falls (Hatsumi Nakagawa); soccer on beach (Evoy Zaniboni); man and farmed salmon (Monica McIsaac). The photograph of Sir Richard Burton is from vol. 2 of Isabel Burton's *Inner Life of Syria, Palestine and the Holy Land* (London: Henry S. King and Co., 1875) and is used courtesy of the UCLA Charles E. Young Research Library, Department of Special Collections.
Map of Brazil: Visutronx
Cover photo: João Zinclar Lima Silva
Author photo: Theo Harvey

The writing of *The End of the River* was generously supported by the International Development Research Centre (Ottawa, Canada, www.idrc.ca).

The publication of *The End of the River* has been generously supported by the Canada Council for the Arts, which last year invested $20.1 million in writing and publishing throughout Canada, by the Ontario Arts Council, by the Government of Ontario through Ontario Book Publishing Tax Credit, by the OMDC Book Fund, an initiative of the Ontario Media Development Corporation, and by the Government of Canada through the Book Publishing Industry Development Program (BPIDP).

ECW PRESS
ecwpress.com

Printed and bound in Canada

For Hatsumi

CONTENTS

PART III: *Velho Chico*

ACKNOWLEDGEMENTS

This book would not have been written without the financial and moral support of Canada's International Development Research Centre (IDRC). Brian Davy, much-travelled mentor and friend, championed the project, as he did so many of my overseas endeavours. Thank you, IDRC.

Many of the experiences that went into this book came about during fisheries projects funded by IDRC, the Canadian International Development Agency, the United Nations Environment Program, the World Bank, the Food and Agriculture Organization of the United Nations, the Vancouver Foundation and many others. To all the project managers I dealt with during those years, I hope the parallel notebooks I kept have not proved too much of a surprise. You made it all possible.

Joachim (Yogi) Carolsfeld, with his ability to walk into any situation and emerge with two airline tickets, first opened my eyes to Brazil and has been the source of countless leads, fruitful ideas and explanations. His large presence lurks behind many pages of this book. My friends and helpers in Brazil are too numerous to list, but I would still like to say a special thank you to Hugo Godinho and his wonderful family, Barbara Johnsen, Arley Ferreira, Carlos Bernardo, Pedro Melo, Norberto dos Santos, Bernardo Sardão, Erika de Castro, João Suassuna, Fátima Perreira, Miguel Ribon, Beth Szilassy, Evoy Zaniboni and Uwe Schulz. João Zinclar and his magic lens showed me what a real photographer can make of the São Francisco and its people. Tom Mace was a great companion in Colombia and Venezuela, where Julio Perez opened many local doors. In the Philippines, Clarissa Marte and

Corazon Duenas patiently explained things.

I hope my friends in various fisheries departments in North America also forgive those parallel notebooks, and I especially thank Gary Logan, Neil Schubert, Devin Bartley, Herb Redekopp, Blair Holtby, Matt Foy, Chris Wood and Ian Bruce for taking me through the many areas that were outside my field. John Fraser continues to be an inspiration as well as a kind of living *Bartlett's Quotations*. The staff at World Fisheries Trust remained indispensable even after I abandoned them to write this book.

In Japan, my father-in-law, Kazuo Nakagawa, answered all my *gaijin* questions and then some, and my mother-in-law, Mitsue, never stopped feeding me. Minoru Kihara has been a faithful friend and colleague and has always done his best to answer my pushy questions about fisheries research in Japan. A special thank you is due to the Toyota Rent-A-Car office in Kushiro, Hokkaido, for introducing me to Miss Tojo.

Three Victoria writers kept me going with encouragement and criticism I could actually use: David Greer, Sarah Harvey and Lynne Van Luven (who deconstructed the manuscript professionally for me, and improved it immeasurably). Jack David at ECW Press believed in the manuscript of a first-time author, and my editor, Emily Schultz, improved it greatly. Thanks also to Crissy Boylan and Rachel Ironstone for all their assistance during production, to publicist Simon Ware and to Heather Sangster for copy editing a manuscript in three languages. Mark Hume and David Anderson read early drafts and provided comments and encouragement, and I am very grateful to Thomas Lovejoy and James MacKinnon for going so far as to endorse it. Theo Harvey did all the electronic manipulations needed to convert my snapshots into publishable photographs.

The photographs interspersed throughout the text are by Brian Harvey except: Rio Negro family (David Greer); sockeye salmon (Carmen Ross); Iguaçu Falls (Hatsumi Nakagawa); soccer on beach (Evoy Zaniboni); man and farmed salmon (Monica McIsaac). The photograph of Sir Richard Burton is from vol. 2 of Isabel Burton's *Inner Life of Syria, Palestine and the Holy Land* (London: Henry S. King and Co., 1875) and is used with permission of the UCLA Charles E. Young Research Library Department of Special Collections.

"He wanted to be the ichthyologist,
but I would rather have been the fish.
His dreams were of capture, mine of escape."

— *Richard Flanagan,* Gould's Book of Fish

A map for a journey

This book has three parts. Two of them — the bookends — are about a Brazilian river I met by chance but that occupied my thoughts for a decade: the São Francisco.

Part One, "Against the Current," introduces rivers in general, the São Francisco in particular, and me — your guide and narrator. Part Three, "Velho Chico," takes you back to the São Francisco, armed with some tools for digging deeper into what's happening there and what it says about human folly and the way we treat nature.

What happens in between? Part Two is the toolbox. It's a collection of stories that introduce the facts of life in science and fisheries, without which the story of the São Francisco would be just another tale of stupidity and environmental woe. To sustain your interest, I've called this part "Science, Sex, Sushi." It's okay to think of it as a digression, because it really does lead somewhere.

Lots of books could be written about the São Francisco; I hope they are. Some might be more linear than this one. But I've told the story this way because it's not just about cubic metres of water and political decisions, it's also about science, fisheries, aquaculture, history — even religion. Above all, it's about people. I'm one of those people, and I've included myself in the story because if you're going to care about this river, I think you should first understand why I do.

A note about names: most of the names of people described in this book are their real ones. I changed only a few: language instructors Lucia Lopes and Jaime Diaz, fishing lobbyist Bob Chambers, and biopirate Andrew. Brazilian names can be rather long; I mostly use the streamlined, everyday version.

SOUTH AMERICA – BRAZIL

ATLANTIC OCEAN

N

VENEZUELA

GUYANA

SURINAM

FRENCH
GUIANA

COLOMBIA

ECUADOR

Manaus

Rio Amazonas

Belém

B R A Z I L

Rio São Francisco

Maceió

P E R U

Salvador

Brasília

BRAZILIAN
HIGHLANDS

BOLIVIA

Belo Horizonte

São
Paulo

PARAGUAY

Rio de Janeiro

CHILE

PACIFIC
OCEAN

Pôrto
Alegre

ARGENTINA

URUGUAY

0   500   1000 km

Teresina   CEARÁ

RIO GRANDE
DO NORTE   Natal

Patos   João
Pessoa
PARAÍBA
Campina
Grande
Brejo da   Recife
Madre
de Deus

Cabrobó

MARANHÃO

PIAUÍ

Juazeiro
Represa
de Itaparica   ALAGOAS
Maceió
Barragem de   Canudos   Penedo
Sobradinho   SERGIPE   Piacabuçu

TOCANTINS

GOIÁS

BAHIA

BRAZILIAN

Salvador

HIGHLANDS   Carinhanha

Ilhéus

Montes Claros

ATLANTIC
OCEAN

Pirapora
Três Marias   Rio das Velhas   MINAS GERAIS

Represa de
Três Marias

0   500 km

SÃO
PAULO

Belo
Horizonte

ESPÍRITO SANTO

São
Paulo

Rio de Janeiro

**SÃO FRANCISCO
BASIN**

Santos

PART I

Against the Current

CHAPTER 1

DIFFERENT LANGUAGES

Lessons from the creator of The World

Say what? Norberto dos Santos on the receiving end of my Portuguese.

I HAVE A JAPANESE FRIEND CALLED MISS TOJO. Maybe "friend" isn't exactly right, although "acquaintance" doesn't seem to capture her either. "Presence" might be best; she certainly is that. We'll meet her properly later.

Miss Tojo is one of the reasons I've never learned Japanese very well, because she never lets me get a word in. Miss T isn't interested in communication. Instead, she hectors, and she does it from inside my head. I hear her all the time now: whenever I hear the words "global warming," every time I read about another fishery closure or another species added to the list we've learned to grow in cages on soybean meal. I think these things are problems; Miss Tojo prefers to view them as novelties, or as obstacles to be overcome with planning and hard work. When I spotted yet another new fish in the supermarket the other day — Russet Perch, I think it was called — Miss Tojo piped up immediately. "Excellent," she said. "What will they think of next?" I had to duck into Meats to shake her.

Miss Tojo is always there. She lurks behind every page of this book, even if you can't hear her yet. Miss T doesn't mind waiting, as long as she gets what she wants. She knows I'll get to her eventually. When she realizes the book takes place mostly in Brazil, she'll be annoyed, there's no denying it, but dealing with Miss Tojo's annoyance is something we're all going to have to learn to do. We might as well start now. When we can't ignore her any longer, I'll do my best to translate what she has to say.

Translation is something I'm getting better at. I do a lot of it in this book. Not just the familiar kind, from one country's language to another's, but also deciphering the jargon of specialists who may be from one's own country but whose pronouncements are incomprehensible. Scientists — *especially* scientists — bureaucrats, environmentalists, sociologists: does anybody know what these people are talking about? Half the time I don't, and I've dabbled at being all of them.

I can't tell you exactly when it happened, but at some point in my career I realized that, as a scientist, I was speaking a language few people understood. How I figured this out is one of the themes of this book; I might never have figured it out at all if I hadn't had the wit to take my training as a fish biologist on the road. But I did travel, and it didn't take me long to realize that if I was going to learn anything about the world beyond my backyard, I would have to pick up a few more of the old-fashioned kind of tongues. So even if Miss Tojo's incessant nagging thwarted my attempts to learn Japanese, I did learn a few others — sort of. Fisheries projects in Southeast Asia (where I went first) weren't much of a challenge because lots of the locals spoke English. But South America, where I went next, was different.

After my first few trips to South America I knew it was no place for a linguistic ignoramus. Two weeks in Venezuela in the care of a loquacious professor was enough to convince me there might be good scientific work done here, but the "in the care of" part was a problem. I might as well have been a Western visitor to the old Soviet Union, with my minder always at my elbow, arranging hotel rooms, checking me into flights, taking me to approved shops. It was infuriating, and the experience in Colombia was worse; falling in love and meeting a *narcotráfico's* veterinarian were fine things to have done and I'll tell you about them later, but they should never have been attempted on a twenty-word vocabulary. There's a fine line between honoured guest and appendage, and when I couldn't get directions, let alone read a map, I felt uncomfortably dependent. What if something went wrong? I couldn't even pick up the phone and order a pizza. Worse, I didn't even know if people *did* that.

So I decided to take Spanish lessons. Victoria, BC, where I live, has a small Spanish-speaking community, but the only lessons these people need are in dealing with a culture in which people keep their voices lowered and never touch one another. There were only two teachers advertising, and only one of those held formal classes. "Learn Spanish Using the Renowned *El Mundo* System," the advertisement said, next to ads for ear coning and discreet Asian escorts. *El Mundo* — The World — I liked the sound of that. I wrote out a cheque.

El Mundo met in the evenings in a nondescript upstairs office in a strip mall; there was a fast-food outlet with terrible coffee below and the actual centre of The World was hard to find. There weren't any signs saying *El Mundo*. I poked my head into the only room with people in it and took my place at a long table ringed with dented metal chairs. It was raining outside, a freezing January drizzle, and I wondered what the weather was like in the Magdalena River or the Orinoco. Probably raining there too, but the rain would be the temperature of tears. Soon, very soon, I would be able to say, "It's raining" in perfect Spanish.

There were plenty of chairs, but only four other people. One of them kept looking at his watch. Finally he stood up and introduced himself as our instructor, placing a thick photocopied manual in front of each of us. "Welcome to *El Mundo*," he said in English. I noticed he had no copies left over; *El Mundo* must be a tight ship. The cover of my manual, which was about the thickness of a weekend newspaper, said, *"El Mundo. Spanish for Beginners. By Jaime Diaz."* I glanced at the young woman opposite me. She was staring at the title, mouthing the words.

Jaime Diaz had thick black hair and crumpled good looks but he seemed tired, as though carrying the weight of *El Mundo* on his shoulders had worn him down. He had a resonant voice with what I later found out was a fine Peruvian accent and he wore — oddly I thought for a Latin, especially one this handsome — a brown cardigan sweater. He spoke perfect English.

"I want to ask, Why are you here?" he said. We all looked at our manuals. "You, sir? Gary, yes?" Jaime nodded at a pale young man with protruding front teeth like a gerbil's, and smiled, I thought a bit uncertainly. The young man seemed an unlikely student to me too. He stared fixedly at the tabletop.

"Me and my friend," Gary finally said. "In Mexico there. We do, um, security work." He cleared his throat. "Down there in Mexico." Jaime Diaz laughed nervously and I tried to imagine the fellow with a badge on his chest and swinging a nightstick. Maybe his friend provided the muscle, while the little man barked orders in confident *El Mundo* Spanish. Jaime moved on.

"And you?" He nodded to the young woman across from me. "Rachel, isn't it?"

Rachel was stunning. She had wavy auburn hair that flew out in a ravishing penumbra, and her smile was unforced and radiant. "For shopping?" she said brightly. "In Costa Rica?" Rachel's sentences ended with question marks. "I'm going down there on vacation and, like, I want to be able to order stuff . . . and stuff?" I couldn't keep my eyes off that smile. "My friend and me," she added, as though to forestall any offers of assistance.

The third student was a middle-aged woman and I have forgotten her reason for wanting to learn Spanish, but it couldn't have been a compelling one because we never saw her again. Then it was my turn. After Gary and Rachel and the nameless middle-aged woman, my own aspirations seemed out of place. I felt the familiar confusion about how to describe my work, what language to use. I could easily have said, "Going down for a little fishing" or "golf tournament" or even something slightly off-centre like "bird watching in the Galapagos," but for some reason I decided to come clean. Maybe there was still a chance to impress Rachel. I cleared my throat.

"I'm a biologist. I work on fish."

Jaime Diaz looked relieved — interested, even. "What kind of fish?" he asked.

"Oh, migratory ones. The kind that swim up the rivers, you know." I could feel everyone's eyes on me. Gary the security man looked up suspiciously.

"Like salmon?" Jaime seemed genuinely intrigued. Perhaps he was starved for stimulation. I wondered about his home life.

"Sort of," I said. "But only in the rivers, not out to sea like salmon. Fish like" — I racked my brain for an example — "like *bagre.*" *Bagre* was a big catfish.

"Oh, I love *bagre. Bagre* is *muy sabroso.*" He smacked his lips so that we could all absorb this first lesson. "I never knew it migrated, though.

Your work is like, conservation?" Jaime and I were getting along.

"Sort of," I said again, adding, "I help the people there to collect samples, that kind of stuff." That should do it, I thought, but Jaime was implacable.

"What kind of samples?" he asked, leaning forward.

"Sperm." I looked away from Rachel and the middle-aged woman. Gary sniggered.

"Oh, yes," said Jaime. He shuffled his copy of *El Mundo* and the first lesson began.

I only spent three sessions in the office building because by that time I was the only student left. Gary said scarcely a word, stumbling repeatedly over "My name is Gary," so that it was clear to everyone, even to him, that he was never going to progress to "Freeze, *cabrón!* Up against the wall. And slowly, if you don't want to lose another of your *cojones.*" He dropped out after the first week. Rachel vanished the week after; all I had learned about her was that she sold belts in the mall. Jaime took the defections in stride, rubbing his hands together and proposing that we meet privately and continue our studies at my own pace.

That suited me fine, and it lowered Jaime's overhead. Over the next three months or so, as the deadline for my next trip to South America loomed, we progressed through the fourteen verb forms, stumbling on the subjunctive, which I have yet to master. I added vocabulary to an expanding list that I mumbled daily at the dinner table, covering half of the page with one hand and eating with the other. Within three weeks Jaime and I were having actual conversations.

This was when the problem began. To converse, you need a topic. When you are just starting out with a language it's extremely hard to spin any exchange beyond a few sentences unless the topic is trivial or the instructor is unusually creative at seizing whatever ideas you give him and tossing them back in ways you can respond to. Even for the creator of *El Mundo* — The World! — this was difficult. Nobody would think of approaching a four-year-old and launching an exchange on the upcoming election, but that's what a language teacher is expected to do. Inevitably, our conversations would veer in the direction of the only thing Jaime and I had in common: we were both men in our early forties. After two or three of these sessions at my kitchen table, the ratio of English to Spanish was actually increasing. We were going backward.

"Como vai?" I would inquire as Jaime settled himself with the day's photocopied lesson and his handwritten corrections on my previous week's essay. "Ah, you know how it is," he would reply in English, spreading his hands and cocking his head ruefully (if nothing else, I was learning Latin American body language). "The wife, the job, what are you gonna do?" Toward the end, just before I left to try out my new language in South America, I had learned a lot about the Latin community in Victoria ("Don't go near that one, she's *loco*") but was more or less teaching myself the language.

But the lessons had worked, better than I thought, although there were still frightening gulfs. A lot depended on who was talking. In Venezuela I had two scientific colleagues, Julio and Luís. Julio was no problem; his false teeth forced him to speak slowly. Luís, on the other hand, spoke extremely fast and never seemed to open his mouth at all and I don't recall understanding a single word. Colombia was better; they pronounce the entire word there, not just selected pieces. But when I returned to Canada I realized I was going to have to spend time in Brazil as well, so I threw myself into the conversion, buying a copy of *501 Portuguese Verbs* to replace the Spanish version and starting a brand-new vocabulary book.

Spanish and Portuguese are not that far apart; they share many of the same words and have an almost identical grammatical structure. Once again I simply ignored the subjunctive tense and all its arcane permutations — if I really needed to say "I should have realized that, had we not taken this road, we might not have become lost" I would rely on eye-rolling and rueful slaps to the head. The real problem was pronunciation. Spanish is like Japanese, each word a string of syllables democratically assigned exactly the same emphasis. No accents, no sing-song, no sounds that don't actually exist in English. But Portuguese is not so sensible; it's flamboyant and baroque, bristling with accents and rhythms and weird, non-English sounds, resonating nasal honks from somewhere high up in the ventricular cavities of the nose and forehead, like the tones of a classically trained singer. For *this* stuff, I needed a conversation partner. And so it all happened again.

LUCIA

If Spanish speakers were thin on the ground in Victoria, Brazilians were almost nonexistent. Nobody offered actual lessons, and it took me some determined digging before I located Lucia Lopes behind a desk in a cut-rate travel agency across town. She was certainly Brazilian, and living in Canada hadn't tempered the way she dressed. She was tiny, but her gestures were large. I wondered how her clients kept their minds on their itineraries.

Lucia and I agreed to meet weekly at the Starbucks next door to her agency. Apparently I would also buy her coffee, but I didn't mind; she only charged me twenty-five dollars a session. She didn't have Jaime Diaz's interest in the problems of migratory fish in South American rivers, but then, how many people in Victoria did? How many people *anywhere?* Lucia made up for her lack of interest in fish by being a champion talker. That was the problem; she would say something like, "And your family, Brian, how are they?" and before I could collect my thoughts and patch together a reply in Portuguese she would be off. "My little girl, so difficult. And my husband . . ." Here she would roll her eyes and take a big, bosom-inflating breath and the men in the coffee shop would shift uneasily in their seats.

Her husband came up a lot. I learned the Portuguese for "military" and "away from home." After three or four sessions Lucia asked me for a job because her boss at the travel agency was "an animal." I decided Lucia and I were finished, that I could look up the Portuguese for "jealous" and "fisticuffs" myself. I was ready to go to Brazil.

I'd learned enough Portuguese to start doing my own research on a story that had been eating at me for years — a story about a river in Lucia's country. Its name was São Francisco. What I didn't know was that I was already in the middle of another kind of story: how I had to re-educate myself in order to make sense of Lucia's river and its problems. The river was a stunning example of the freight train of human folly, thundering driverless toward the inevitable hairpin curve in the track. But it already had its champions. Should I — *could* I — join them? Answering that question meant putting both

Brazilian and personal events in the larger context of history and science, and telling them as I experienced them — as bits gleaned here and there in the course of a personal journey that meandered as much as the river.

This book is about those journeys: how the São Francisco got to where it is, how I stumbled into it, how I let the river carry me along until it became clear that I was taking a journey without a destination. In the end, my struggle to stay afloat in the São Francisco story forced me to rethink what I'd been trained to do — which brings me back to Miss Tojo.

I bring up Miss Tojo again because we all search for explanations, for unifying images we can hang our hats on, for anything to help us answer the awful question, "How the heck did we end up like *this?*" On the personal side of things, I don't think Miss Tojo is much help. But for the mess that particular river was in — and not just that river, but other rivers, lakes and oceans, and everything that swims and crawls and waves to and fro within them, out of sight and minding its own business — the image of Miss Tojo is the one that sticks in my mind. You'll see why when we meet her. But first we have to get through a lot of other stuff.

CHAPTER 2

EXPECTING THE
WORST

Going back to Brazil

Chaos and destruction: a charcoal truck in Minas Gerais, Brazil.

I DON'T TRAVEL WITHOUT DRUGS. Not long distance, anyway, not between continents. There's nothing romantic about waking up at two in the morning in a far-off place, staring at the ceiling, reviewing a long list of flimsy reasons for being ten thousand kilometres from home, especially when you're expected to bound out of bed at six-thirty like everyone else: shower, shave, eat, perform. I discovered the benzodiazepine group of tranquilizer twenty years ago and have used them as knockout drops ever since. Mother's little helper, and only when I travel, but fifteen milligrams a night for the first three days of a trip keeps me down for eight good hours. So much for jet lag.

Periodically I have to top up, visit my supplier. Not often; a prescription of thirty tablets lasts for years. Between trips, it sits in the bathroom cabinet, absorbing moisture from the shower and communicating chemically with neighbouring bottles of hairspray and mouthwash, silently transforming into something I should be flushing down the toilet, not experimenting with in a strange hotel room at the end of a twenty-four-hour trip. Usually, though, I forget to check the bottle until the last minute.

That's why I was in my doctor's office now. The next day, I would leave for Brazil again, my third trip in as many years, and I only had three of the magic yellow pills left. These ones weren't even yellow anymore. Some of them were turning white, and when I shook the bottle little pieces fell off. I might not sleep if I took one of those things. Or I might not wake up.

My doctor's waiting area is an odd shape, more hallway than room. A half-dozen chairs line one wall across from a low table scattered with copies of *Chatelaine* and *Sports Illustrated* that I imagine to be crawling with some antibiotic-resistant bacterium unknown to medical science. You don't actually see the other patients because they're beside you, a curious arrangement that precludes conversation. I sat down with a bundle of papers about the São Francisco River, knowing there would be time to kill. That river was where I was going — again — and grinding through reports on its many tribulations had become a big part of my job.

These reports came from all directions — government flacks, academics, non-governmental organizations (NGOS) — and could fill almost any amount of time. More than once I'd got halfway through one of them, cursing the Brazilian addiction to flowery language and Dickensian sentence structure (Lucia never talked like this!) before realizing I'd read the damn thing already. If "read" is the right word. My Portuguese was adequate now, but there were still times when comprehension veered suddenly, leapt a guardrail, exploded into flames. Actually, there were many such times.

But I still had to read about this river, because what was unfolding there was a story that wouldn't go away, played out on a stage of inhospitable beauty and as old as human history: the fight for water. The São Francisco had crept up on me over two decades of my studying fish and fisheries, travelling around the world and absorbing lessons in the frailty of natural ecosystems and the perverse determination of humans to push them to the limit. For twenty years I had assumed that, as a scientist, I knew something about aquatic ecosystems and that, as a conservationist, I knew how to defend them. Then along came the São Francisco to turn everything on its head. Being in a doctor's office made a kind of sense, because this São Francisco River was like an artery: one that was collapsing. And now some people were going to stick a very big needle into it, and what they said they were going to do with the blood just didn't make sense to me. Even worse: if the reports I was slogging through were right, what they *said* was a very long way from what was really going to happen. The river was a mess.

I sat down and snapped the rubber band off the day's stack. On my left was a middle-aged woman in a shapeless sweatshirt and pink running shoes, with a small bundle of clothes sitting next to her. Two chairs away, on my right, an elderly silver-haired man in a dark suit was absorbed in a single sheet of paper he held at arm's-length in front of him. The man wore heavy black horn-rims, and the sheet of paper seemed to be covered in columns of figures. I opened the first document, printed that morning from an e-mail sent from a friend in Brazil. This one appeared to be about a national plebiscite on the plan to divert the São Francisco River — to stick the needle in and pull back on the plunger. Almost immediately, I hit a verb I wasn't sure about. Did "prever" mean foresee (I thought it did), or did it convey more of a sense of "preview?" An important distinction: one meaning suggested there was actually some substance to the idea of a plebiscite, the other that it could just as well be hot air. I began to turn this over in my mind when the middle-aged woman next to me suddenly spoke up.

"We're going over to Europe next summer. To *Europe.*" Her voice was really quite loud. The little bundle of clothes said, "Oh, yes." I skipped to the next sentence, looking for clues to the meaning of "prever." I wished I had brought my dictionary.

"We're going with Tom and Verna!" shouted the woman.

I skipped again, to something about a "deliberative session" that would take place, when? I looked at the calendar on the wall. A month ago; so much for that. Further on, the State of Pernambuco had just granted $95 million reais to start building two canals and a couple of dams. This was important stuff: evidence that, despite the massive opposition to the diversion project, concrete was about to be poured. I needed to know about this. I tried to tune out the canned music in the waiting room. My doctor always seems to be playing Schubert, and I love Schubert. The trouble is, he's more compelling than a polemic in Portuguese. The bundle of clothes spoke.

"I don't know, dear," it said in a tiny, querulous voice. "Vera? I just don't know . . ."

"Not Vera, Mum, *Verna!* Her and me have been friends all our *lives,* don't you remember? She married that Tom after the other one took off. Come on, now."

They got to their feet and followed the receptionist down the hall to an examining room, the mother clinging to her daughter's hand. The old lady was bent, wearing running shoes that looked much too big for her. Maybe she was seeing the doctor for her edema. The door to the examining room closed, and I turned back to the problems of the river. A voice came from inside the examining room, only slightly muffled: "Well, *I* wouldn't have married him, that's for sure." I gave up, put my São Francisco papers down and wondered if the flayed-looking *Sports Illustrated* on the table was safe to touch. Suddenly the man on my right spoke.

"Hello there, neighbour," he said.

I snuck a look down the line of chairs. He was staring intently at his piece of paper.

"My name's Charles Galvin and I sign all the cheques around here. If I don't sign, nobody gets paid." He had a forceful, resonant voice, with a bit of Gomer Pyle goofiness around the endings of the words.

"I have the same problem," I said.

"Sue them," he said vigorously. "For every last penny." And then, "I was the secretary of the Legion chapter, you know. Chapter 131. In Orillia." He turned the sheet of paper over and frowned at the other side, which was blank.

"I've never been to Orillia," I said. I turned in my seat to help the conversation along, except that it wasn't a conversation. Charles Galvin was talking to his sheet of paper. He opened his mouth again.

"Orillia won't thrill ya." He stared grimly at his columns of numbers. The woman and her mother shuffled out of the examining room and retrieved their coats from the rack, the old lady standing silently while her daughter wrestled the garment on her. I got to my feet and nodded at Mr. Galvin, who was shaking his head.

"I can't sign this, you know," he called after me.

I got my supply of brand-new, shiny yellow pills, enough for the next few trips to Brazil. When I left, there were more bodies lined up against the wall, and the receptionist was attempting to separate Mr. Galvin from his ledger and his curious train of thought. Or thoughts, rather: each one consistent and clearly expressed, but with no connection between it and the next that I could see. Linear but disconnected; no pattern, no sense. It was like the trip to Brazil I would take the next day. I was returning to a river, a place I thought to be important, a place of beauty and conflict — but what would happen next? I'd been going to far-off places for decades, and the only unifying concepts seemed to be that I was a biologist, that I knew something about fish, that the projects I had dreamed up over the years seemed to interest people, and that I somehow managed to get paid to do them. If there was a principle, I wasn't sure what it was; if it was just that I was a fish biologist with conservationist tendencies, that concept seemed to be about as insubstantial as Mr. Galvin's list of numbers. His life wasn't going anywhere. It had been derailed, like his thoughts.

Tomorrow I would be on another plane; the day after tomorrow I would be standing in the blistering sun beside a big, brown river that I knew to be important, knew to be threatened. It had an unequivocal place in the trains of thought of millions of Brazilians. I just wasn't sure where it fit in mine. Maybe, this time, things would be clearer. I put my hand on the doorknob.

"A bunch of vultures, all of them," said Mr. Galvin.

Calling Mr. Fix-it

I've heard of travellers who can't wait to be on their way, but I don't slip the ties easily. Always the same countdown: visas that arrive at the last minute; sheaves of slippery traveller's cheques to be signed and squirrelled away in strategic and immediately forgotten hiding places; distracted and desperate shopping for gifts, like a grim Japanese tourist; scraps of paper with lists of essentials omitted the last time; itineraries copied and relatives reassured. Instead of anticipation and excitement, I succumb to a sense of dislocation and doom that's made all the worse for being the kind of shameful secret no seasoned traveller would ever admit to. Even if there exists such a thing as the ability to jump lightly from one's own surroundings and alight in another land, I don't have it. The day before departure, I find portents everywhere.

The night before I left for Brazil with my shiny new pills, I walked the dog. It was a cold November, the day's steady rain relenting now, but the sky still wet and the wind snatching dead twigs off trees and flinging them in my face. I walked fast, Bonnie nervous in the storm and tugging me along. What a luxury it was to be able to walk without my eyes glued to the pavement, scanning for the crack or cavity that would catch my foot like a wrestler's hand, toppling me face down into a tangle of rebar. In twenty-four hours I would be in a place where sidewalks were hummocked and booby-trapped. This didn't make me feel any better.

I knew this trip to Brazil was important, but I'd been on too many that weren't. My life was beginning to feel predictable, like a bad airline movie watched with the sound off. Maybe the movie and the trips were someone else's life, not mine. Maybe the president of World Fisheries Trust was somebody else, not me. That person had come back from Rome only a few months before, from what should have been a stimulating three days of fisheries meetings at the Food and Agriculture Organization of the United Nations (no wonder they shorten it to FAO), with characters that are supposed to be one of the best reasons for travelling: the Indian scientist whose nose hooked almost to his upper lip (how did he *shave?*), the yellow-toothed Brit who rubbed raccoon eyes and said, "Sorry, just flew in from Burkina Faso." For a novelist, these people would have been gold.

I'd slept poorly in Rome, waking once from a dream in which I'd got hopelessly lost in FAO's warren of identical corridors and blind alleys. What woke me was a summer storm of biblical fury: thunder and lightning and curtains of rain. I got out of bed and opened the door to the courtyard, where the palm fronds glistened under the arc lights on the neighbouring balcony and the Italian flag hung, sodden and limp. This was *Rome*. I shouldn't be hating this trip. The people were wonderful, even if they weren't much like me. I just had to stop trying to live their lives and start living my own. The cool rain plastered my hair and bounced off my feet, and I thought that maybe if I stood there long enough something would wash away. The meetings themselves had been mostly posturing, those exotic experts marking out their territories like cats pissing in their owners' backyards. And to my disgust, I found myself doing it too. Smile, lift a leg, dribble — it was the only way to survive in this business.

And now I was going back to Brazil. I rose before dawn for the wordless ride to the airport. Finally, detachment began for real: the bag with all my hastily crammed papers and presents and all-purpose clothing vanished through the vinyl curtains behind the check-in counter. Security stroked my laptop with Q-tips and pronounced it fit to travel, and everything I'd set in motion coalesced into a single image: the departure gate. I waited with the people taking regional flights to comfortable places like Prince George and Kamloops: a day or so, then back home for the weekend. An Air Canada agent announced, for the third time, "Once again, this is the final call for Toronto." Four Brazilians straggled toward the gate, laughing and punching one another in the shoulder. They weren't in any hurry; they knew how long the flight was.

Miraculously, I was granted an upgrade between Vancouver and Toronto. Maybe the agent felt sorry for me or maybe he just made a mistake, but it was still like winning the lottery. I settled into my throne, rearranging piles of blankets, bottles of water, menus and pillows. If I had to go to a flaming death in Brazilian traffic or an ignominious one under an overturned boat on the muddy São Francisco, I might as well go full of chocolate truffles. I looked around at my fellow executives. Across from me was an ordinary-looking fellow in his early forties, in jeans, pullover and suede shoes. He looked tired too.

"Sixteen-hour days," I heard him say to an excitable man in the row ahead of him, a small person, twisted around so that his head just peeked over the top of his seat. "Sixteen-hour *days*, man. Brutal." The little man nodded ecstatically, and the flight attendant brought champagne. There was something about the object of their attention I couldn't place, as though he were someone mildly famous, an actor in a commercial perhaps, an image glimpsed and retained and now, as we gained altitude, bothering me intensely. I had a bag full of reading to get through, articles and reports about the controversy over the São Francisco River diversion, and my stay in Brazil would start with a bang. I really needed to get to work.

But I kept sneaking looks. The guy had sandy hair, almost orange, buzzed close to the scalp to accentuate a ponderous jaw and full lips. With a corncob pipe he might have been Popeye, and the hands were certainly enormous, with fingers the size and colour of wieners long boiled. One wrist bore an understated metal wristwatch; the other, a slender double band of metal, like brushed aluminum. He wore domed earphones he must have brought aboard himself, and the big fingers leafed fastidiously through a magazine on home improvement. His fingernails were immaculate. The sense of having seen him before wouldn't go away. Football player? Too old. A politician, perhaps — but a politician would be pecking at his BlackBerry or skimming over the kind of reports written by people like me. This guy just turned the pages of *Canadian Home and Style*, tired-looking but serene. Just another sixteen-hour day.

We flew on, over the prairies, over the endless lakes of northern Manitoba and Ontario already icing up for the winter. In twelve hours the view from the plane would be green, with the occasional shimmering snake of a river, but now it was all brown and pewter. The flight attendant came and went. She was coolly efficient with me; I guess she knew an imposter when she saw one. Finally, as the brown fields beneath us rolled up toward the shores of Lake Ontario and we closed in on Toronto, she sank to her knees in the aisle beside my mysterious neighbour and began to question him. He seemed to be listening carefully, although most of what she said was drowned out by the vacuum-cleaner rush of the engines. But I could hear his reply.

"What you want is an engineered floor, not a laminate. It's got all the advantages of a composite, but there's still a good three-eighths of solid wood on the top." The flight attendant nodded rapidly, and her pony-tail bobbed. She gripped the arm of his throne and looked up.

"See, with your engineered floor you get the choice of any kind of wood. There's really no downside. Oak, cherry, even pine if you want it." He raised a meaty finger. "But pine, I don't recommend it. Too soft. You from around here?" He gestured out the window, which was filled with nasty grey clouds. We were going down through them already. The flight attendant should have been collecting wine-stained copies of the *Financial Post* and scurrying to her crash-proof seat. Who *was* he? Some synapse in my brain sputtered and sizzled and I reached for the in-flight magazine. I riffled to the full-page drawing, in comic-book style, of an action hero in coveralls and a sleeveless undershirt swinging from a lamp cord. I knew I'd seen him! In the other arm he held a flailing woman with a look of astonishment on her face while, down below, a suburban home collapsed in a cloud of smoke. It was definitely the same face, impassive and competent under the buzz cut. "She called for a contractor and what she got was a *hero!*" said a balloon under one of his raised workboots and, under the other, "Watch the mighty Mike Holmes rescue real homeowners from contracting disasters!" Exploding yellow letters across the top said, "Holmes on Homes."

"I live in Montreal," the flight attendant said.

"Never been to Montreal," said Mike Holmes affably. "But I hear it's beautiful."

"Actually, I'm originally from Europe."

"Oh, well now, Europe." He pronounced it *yer up*, like an umpire. "I've been to Europe. It sure is beautiful too."

"And my name is Marina," she said, getting to her feet and straightening her uniform. "I'll remember about that floor, what was it?"

"Engineered floor, that's what you want. Resist the laminate."

We landed. Mike Holmes returned to his copy of *Home and Garden*. His right knee jiggled. When we had rolled to a stop he slipped his grey blazer on and disappeared ahead of me up the Jetway. Several middle-aged men fell in step with him, making sawing and hammering motions with their hands.

Here was a man who specialized in tearing out contractors' mistakes, in fixing messes. He was wildly popular. Should I be running ahead, getting his card, taking it with me to Brazil? Surely that's what the country would need in five or ten years, when the army of contractors had left the northeast and President Lula's fantastic network of concrete canals full of water from the São Francisco started to develop hairline cracks, when the pumps seized and the national debt had taken a dizzying spike as a result of the mess.

I imagined Mike flying calmly down to Brazil, his earphones on and his sure hands working steadily through sheaves of engineering drawings. Switching flights in São Paulo, sleepless but determined, changing to his overalls in the first-class lavatory and alighting finally on a blasted plain in Pernambuco to scowl at the clogged concrete intestines radiating away into the heat-shimmering distance. Despite the ferocious sun, his sandy crewcut would bristle unprotected. His expression would be grim.

"These *adutores*," he would say, flinging his arm toward the horizon, "they gotta go. All they do is burn off water. All of 'em, just rip 'em out." I imagined the canals flying off into the *sertão*, comic-book style. "And those *bombas* back there, ditto. No mercy. Whaddya want pumps for? You've got water falling out of the sky, so use it already! Leave the poor river alone." Lula's engineers would hover anxiously, avoiding his gaze. "You paid, what, five *billion* for this? When all you gotta do is fix up all those reservoirs you built ten years ago?" The engineers would shrug and look down.

"Incredible," Holmes would growl, scratching his burning scalp. "This country must have cement companies up the ying-yang."

CHAPTER 3

THE SÃO FRANCISCO

A toe in the water

Friendly town: boy on the promenade along the São Francisco River, at Pirapora.

ALL TRAVELLERS HAVE TO COME FROM SOMEWHERE, and that somewhere inevitably colours their reactions to a new place. In my case, the frame of reference for the São Francisco is another river, the Fraser. If you're from southern British Columbia, the Fraser is inescapable; the river barrels around the Coast Range Mountains, flows right past Vancouver and empties into the Pacific near the international airport. You can't miss it. The Fraser and its many tributaries once supported the greatest salmon fishery in the world, and life in coastal southwest British Columbia is still deeply coloured by that unlucky fish — or rather, by its memory, because the commercial fishery in the Fraser is hit and miss now.

Like the São Francisco, the Fraser was a river of discovery and exploration. As did many Brazilians, I too grew up with stories of the glorious exploits of men who opened up a country by following a river. Simon Fraser, the British explorer who bushwhacked his way through much of British Columbia in the early nineteenth century, may not have been a saint like St. Francis, but sometimes it seems that half of Vancouver is named after him. The human record on the two rivers has similar stains too. Both were home to ancient civilizations, although the aboriginal peoples on the Fraser have fared better than most Brazilian Indians: they're still alive.

The Fraser and the São Francisco share many problems — generally the ones common to all rivers — with the major exception that the Fraser mainstem was never dammed. And the Fraser is shorter, less than half the length of the São Francisco, which leads me to think the word "mighty" is overused. They can't all be mighty, can they? It doesn't matter; the not-quite-mighty Fraser always seems mighty to me because it's my point of departure. Its tributaries flow in like the veins of an alder leaf collecting at the stem. The São Francisco drainage is like a leaf too, maybe a mango instead of the alder I grew up with. For me the two rivers are like trees on opposite sides of a mountain: far apart but stirred by the same winds.

FLASHBACK: THE FIRST TIME I SAW THE SÃO FRANCISCO

The trip to Brazil in which I flew next to Mike Holmes was not my first. It had been preceded by many others, and I first saw the São Francisco River itself in the 1990s. It was a side trip, a stop along the way to somewhere else. I can't even remember why I went there — probably something to do with one of the seven dams the river is afflicted with. Dams affect fish, a point I was good at making to funders, and a message I knew to keep simple. On that first trip, I arrived at the river through the town of Três Marias, because it's the easiest point of access along the upper section of the river's nearly three thousand kilometres. Of the trip to Três Marias I remember little. I can only reconstruct impressions from later trips, but I must have come by private car from Belo Horizonte because back then my Portuguese was hopeless for taking the bus or renting a car. This was before the language lessons with Lucia Lopes, who would still have been in Rio, shedding husbands and dreaming of a better life in Canada.

Now, when I visit that part of Brazil, I stay first in Belo Horizonte with Hugo Godinho, a fish-biologist friend. His house is quiet and cool, with ornate wooden chairs around a dining table where Ica, Hugo's wife, insists on organizing a comprehensive breakfast for me every morning. There's a small courtyard out back, patrolled — very slowly — by a pet tortoise that moves like an arthritic ballet dancer, each leg raised high and brought down *en pointe*. The courtyard is paved in black Minas stone, still warm under bare feet an hour after the sun has gone down. I wake up to sparrows, the neighbour's yappy dog, a rooster somewhere, kids and dishes, the maid humming and the swishing sound of her flip-flops as she goes back and forth between kitchen and dining room, laying out my breakfast of yogurt and black coffee and fresh bread just bought at the end of the street. The neighbours are very close, but high walls and hedges mean all you see are their roofs and satellite dishes.

I usually spend a day or so at Hugo's, decompressing, getting used to finicky toilets and electrical shower heads that are warm only at a trickle. At night the occasional truck cutting through the cobbled streets shakes the house so the iron window grille rattles. If there's a soccer game in Mineiro stadium a kilometre away, I can hear the honking and the small planes circling, the fireworks and, when a goal is scored, the feral roar from a hundred thousand throats. At night we eat out, usually at an open-air place in the neighbourhood, consuming beer and fatty pizza and watching little children scamper between the packed tables, collecting pats on the head. Hugo has the collegial Brazilian way with waiters, bantering, touching on the elbow. Maybe some TV after that, surrounded by Hugo's collection of beer cans. Sometimes we go to his daughter's house for a *churrasco,* a barbecue in the back courtyard that involves more or less continuous snacking on slivers of beef and chicken and cinnamoned pineapple grilled by various sons or sons-in-law. The kids play through it all; when they get tired they curl up on the tiles and go to sleep.

On that first trip to the São Francisco, I didn't have the benefit of such a pleasant layover. But I saw a lot during the five-hour drive north: the outskirts of Belo Horizonte, a maze of tire shops studded with hubcaps, a dirty concrete canal, crumbling apartments with people leaning out the windows for a breath of air, rail-thin men pushing carts and orange-coveralled women shoving garbage into plastic bags. Everywhere the stench of diesel and burning brakes. In the city, billboards with the names of conglomerates like characters in *Lord of the Rings:* Brasremig, Magraf, Medplus. Farther out, sides of pleated hills that seem to have been carved away, patches of bananas and corn, a few cows, ponds that might have fish in them. As we drove farther I would have seen what looked like tile or brick factories and, when the countryside opened up again, the insect version: rust-coloured termite mounds rising out of the ground, lumpy and tall as a man, like the remains of a mud fight between giants.

Fire-breathing buildings perched on piles of slag — what were those? The hacked-down *cerrado* stretched in all directions and the potholes were as big as the car we were in. I especially remember the trucks: great Mercedes and Volvo flatbeds piled with propane tanks or bottled water or gravel, their immense tires red with dust and rumbling level with my eyes and, when you looked up, the brown anonymous elbow of the driver. Other trucks were loaded with blackened burlap sacks piled six metres into the air like a great over-risen loaf of bread, and they roared like prairie schooners from hell. What were *they?* Faded stickers on their rear bumpers read, "Let's Live Without Violence" and "100% Jesus" and, my favourite, "Preserve Nature." On the outskirts of every town we passed roadside stands that were no more than crude shelters of sticks and dirty tarp, with racks holding bottles of honey, cane liquor *(pinga),* hot *malagueta* peppers.

A charcoal truck. The sign says, "Preserve Nature."

Trucks, refineries, *cerrado,* potholes: I wrote them all down in my notebook on that first visit and the writing is an old man's scrawl, snatching at impressions without understanding. Just another endurance test on a bad road. The cause and effect connection between them, and their larger meaning for the life of the river I was about to see, were over my head. Três Marias was another place to deliver what had become a potted message about preserving vanishing fish species, to piss again on my own little piece of scientific territory. The sad truth was that people would listen respectfully no matter how spectacular my ignorance of local conditions. Tour the place, have a fuss made over me, give the well-worn lecture and catch the next ride back to Belo Horizonte. It was like travelling in a bubble, cushioned by interpreters who'd been outside long enough to speak good English. I accepted whatever these affable people laid on, having the certainty of a plane ticket home.

We didn't arrive in Três Marias until dusk that first time so all I could see of the reservoir behind the dam was scalloped edges fringed with bare earth, the tabletop mountains just a black outline in the distance. What I knew about the river itself was this: it was big, it had lots of dams and it flowed north. This last detail, I remember, puzzled me: shouldn't a river flow south, or at least east-west? In British Columbia, the big ones did. The Fraser River starts in a high plateau and winds south to the ocean at Vancouver. That was my norm, so what was all this nonsense about flowing north? Later, when I read explorer Sir Richard Burton's record of his trip down the river in 1867, I felt foolish. Ever the practical man, Burton pointed out that this direction is more pleasant to the traveller going downstream, who need not "catch the sun." Sir Richard was always opening my eyes like this.

I saw the reservoir properly when I woke up. My hosts had put me in a hotel the power company operates not far from the dam. It was a modest place, quiet during the week. My room had a parquet floor, each piece of wood cupped at the edges from heat and humidity. The inflatable toilet seat sighed when I sat on it and the electric shower zapped me every time I touched the tap. But I could take my breakfast to one of the plastic tables out front and look past a row of palm trees and out over the reservoir; that was pleasant. It

*Dourado in the local market,
near Três Marias.*

was low water, just before the big rains, and there was a shelving, a thirty-metre ring of reddish earth around the edge, as far as I could see, as though the reservoir were an enormous dirty bathtub. The dam was discreetly hidden from view, behind barbed wire, and I could just make out the line of floats that marked the no-go zone upstream of it. The actual town of Três Marias was five minutes away, past the fish culture research station where biologists studied the reservoir and dreamed up ways of breeding fish to replace the ones the dam destroyed. None of this sank in, that first time, my notes mentioning only the swollen trunks of palm trees painted white near the ground; the horses that wandered in through the parking lot to crop at the grass; a hummingbird.

Of the town of Três Marias I remember this: chaotic, streets running here and there with no discernible plan; not a decent colonial building in sight; beauty salons; bars with pool tables; sidewalks that petered out suddenly. The bars looked inviting. Most of them were tiled inside, with a cooler for drinks, *salgados* (snacks) like cheese breads and pastries in a display case on the counter, a few plastic tables and posters for Kaiser beer. The posters featured pneumatic girls not actually involved in beer drinking (in South American beer ads it's the men who do that). There was movement everywhere, as though Três Marias just couldn't sit still: shirtless guys on motorbikes; cars with deafening radios pouring Brazilian country music into the air; horses pulling carts laden with canisters of milk and bottles of propane; women who seemed to be wearing their younger sister's clothes. Looking up, I saw tangles of electric wiring, TV aerials strung with spiderwebs, butterflies and, above it all, the circling vultures.

Três Marias wasn't even close to the river, which turned out to be unprepossessing, a sluggish, muddy expanse between caked banks and trees I couldn't identify. The elderly dam backed the river up into the ugly, endless reservoir; below the dam the river chugged downstream (north!) around a bend and out of sight. Some kind of factory at the water's edge, something to do with zinc, a bridge to the other side that shuddered every time one of those top-heavy prairie schooners rumbled over. At a shabby market I dutifully photographed the strange-looking fish I was supposed to be saving. The men digging the creatures out of ice-filled Styrofoam chests were leathery and cowboy-hatted and I couldn't understand a word they said. I didn't even learn their names — the men or the fish. I remember swinging my camera to shoot a man leaning out of his disintegrating car, his face out of a Brueghel painting and burned the enduring ochre of a fair-skinned person forced to spend a life in the sun. Two small children craned halfway out of the corroded windows, clowning. The parts of the car that weren't perforated by rust were caked with red dust. The road we were on, someone told me, went all the way to Brasilia. But that was in the days before I looked at maps, so it didn't mean much. I had some greasy fish at a restaurant beneath one end of the bridge and watched a car come to a stop right in the middle, two guys jumping out to push. If one of those big trucks came along, the sky would suddenly be full of those black burlap sacks. I hoped they wouldn't land on me.

That, to my shame, was it. A dirty river, catfish in a disintegrating cooler, inexplicable people. Três Marias might have been the gateway to the upper São Francisco but there wasn't a tourist in sight — why should there be, when Brazil had the Pantanal and the Amazon? I had seen a kilometre-long section of a three-thousand-kilometre river and what impression could I bring back from that? It was like picking a single gene out of someone's DNA and expecting to know them. I don't remember giving my talk, but it couldn't have been inspiring. Maybe I bought a T-shirt in one of the cut-rate clothing stores, or a brick of cheap local *goiabada* (guava jelly) to take home and put on my toast. Probably I caught a flight home from Belo Horizonte after retracing my steps: more trucks, more refineries, more *cerrado,* more potholes. They looked just the same on the way back, although my notes tell me that one especially big pothole was stuffed with branches and that one of those top-heavy trucks was overturned by the side of the highway, surrounded by a knot of men. It looked like a dead hippo.

BARBARA

If I hadn't met Barbara Johnsen just before I left Três Marias, I would probably never have returned. Been there, done that. I already knew more about the São Francisco than nearly everybody in North America, and there were other places to go.

But I did meet Barbara, even if it was only by chance. One of the organizers thought I might be interested in how the town dealt with environmental issues — why not? Barbara's office was in the *prefeitura,* city hall, up a steep side street across from a hole-in-the-wall, aquamarine *sorveteria* that seemed to have a hundred varieties of ice cream. Her tiny metal desk was dwarfed by a three-metre tall maquette of a florid lady in a bright green dress. Barbara herself was almost as imposing. Everything about her flowed: long loose dresses, unfettered flaxen hair, even her speech, so unusual I thought at first she must be — what did I know about this place? — drunk. Her English came sliding out slurred and slow, in a voice deepened by the cigarettes she kept fumbling for in her leather bag. Her eyes were a startling Nordic blue: a chain-smoking Brünhilde in the heart of Brazil.

"You like her?" Barbara waved a cigarette at the looming paper lady. Apparently you could smoke in an office here. I hoped the statue wouldn't catch fire. "Her name is Maria. I make her for the

Barbara Johnsen and one of her three Marias.

children, for the festival of the river. Maria is the one Maria, and we have the two others. So, Três Marias."

Festival? In this place? Barbara shot smoke past my ear. "I like her," she said, cocking her head and looking up at Maria the First in a way so curiously free of pride that it seemed she must be referring to someone else's creation. She was smiling; they both were. I wondered if Maria didn't have something of the *Mae d'Agua* in her, the water spirit that Sir Richard Burton mentions as inhabiting Brazilian rivers, a malevolent Lorelei with hair like threads of gold. Burton speculates that the origin of the *Mae d'Agua* may have been the large otter that used to be common in the river but had the misfortune of having to compete for fish. Hunters would plug the breathing holes in its tunnels in the bank, opening the entrance later to kill the drowning otters. Maybe they really did come back as evil spirits.

"She's beautiful," I said and meant it.

"Yes," said Barbara. "Now let's go and see the river. We can walk." I followed her flowing skirts out into the furnace of the street and downhill, over craters in the sidewalk, past butchers and hair salons and what seemed like far too many shoe stores for a town this size, down a narrow cobbled street with modest houses behind iron gates, their tiled interiors open to whatever breeze could penetrate. "My house," Barbara said without stopping, waving at a particularly small one tucked in behind a small forest of flowering bushes, before we reached a levelling off and a narrow bridge. A pregnant dog panted on its side in the shade.

"Stand here," Barbara said, positioning me at one corner of the bridge out of the way of the cars and motorcycles shooting across it. "Down there is the river." I could see nothing, only a gulch with houses on either side. "Over there, the zinc refinery." She waved, and I thought that must have been the "factory" I saw. "They add their bad things to the river, and so do we. Look there." Now she was pointing down the gully, at a white plastic pipe jutting out from one of the houses. Actually, most of the houses had the pipes, which ended in midair above a strangled-looking stream. As I watched, a little blurt of something came out of one of the pipes, heading for the river that Burton drank from for four months, water he described as "a transparent green, like the mighty Zaire." Now it was a sewage dump.

"Maria helps us teach the children about these things," said Barbara. Then, "I will show you the real Três Marias."

I've met zealots, have even tried being one, but I don't like them much. Zealots care too narrowly; they push their way into your space until you yield. If Barbara was a zealot about the river she was unlike any I had ever met. She had information, but it was only there if I wanted it. Despite myself, I began to listen. Barbara rounded up a driver from the *prefeitura,* an ex-fisherman called Carlão ("big Carlos") and we headed out of town. Carlão honked at a pretty girl, then another. *"Minha prima."* He grinned. My niece. Três Marias was full of nieces, it seemed. We crossed the highway, passed a tiny shrine where Carlão stopped honking long enough to cross himself, then headed up a hill to a grassy place above the hotel where I was staying.

"Três Marias was supposed to be here," Barbara said.

"Here?"

"It's sad, yes?" She took in the ruins with a sweep of one hand.

"See, that was a hotel. And there, houses." She walked over to a bush and snapped off a yellow fruit. *"Jambó,"* she said, and held it under my nose. *Jambó* smelled like soap. "I call it grandmother smell." Barbara laughed. Later, as I got to know her, I would see this behaviour time and again. The things that pained her and the things she found beautiful were two faces of the same coin, and the coin was constantly turning. She was right about the site; there were flat overgrown spaces where you could see the traces of foundations. Nothing colonial, more like the footprints of houses that had been razed to make way for something else, except that nothing else had been built here. It was higher here, and cooler than in Três Marias. On a satellite image the ghost settlement looks eerie, as though the buildings were rubbed out by a giant eraser.

"What happened?" I got the feeling there was more to this town than met the eye, that Barbara was in some way a part of that history.

"You know BR-040?"

BR-040 was the highway to Brasilia. Trucks, refineries, *cerrado,* potholes: maybe Barbara could explain them to me. BR-040 crossed the São Francisco. I'd stood on the bridge the day before, the dam on one side and the refinery on the other and those infernal black trucks making the deck vibrate as though the whole thing were about to fly apart and drop into the river.

"I remember the bridge," I said.

"So you have seen the shit grass."

Was this another topic? I shook my head.

"The shit grass grows under the bridge. Where the shit from Três Marias comes. I will show you another time."

It was hard to keep Barbara on one topic at a time. I wrote down "shit grass" and wondered what it looked like. "But what about *this* place?" I asked.

"JK built BR-040 to connect with Brasilia. In the fifties. When the road crossed the São Francisco here there was nothing, only the old village of Andrequicé in the hills. I will show you that too." She waved in another direction. It all looked the same to me.

"Andrequicé was founded in 1730, in the time of the *capitânias.* All the land belonged to the big *fazendeiros.* But here, there was nobody. So when they decided to build the dam, they did everything at once. Not just the dam, but the refinery too, because the power and the water would be cheap. Everything at once. The workers came from everywhere: the northeast, the places that would be flooded by the reservoir, farther south in Minas."

I scribbled it all down. JK was Juscelino Kubitschek, the president who built Brasilia. That, at least, I knew. *Fazendeiros* and *capitânias* were something to do with land; I could look those up later. "So they built Três Marias for the workers," I said.

Barbara waggled a finger at me, a Brazilian gesture that means "wrong, don't go there."

"No. They built this." She stamped her foot on a foot-high sill of crumbling concrete. "The workers lived here, for four years, there were shifts day and night, it was like an ant hill."

"And when the dam was completed?"

"By then, JK was finished. The military was in power. The company knocked it all down. She kicked the wall again. "So the workers moved down and built Três Marias. The company kept the good land, up here. Both companies, really, because the refinery has its own town, for its workers. Very beautiful, you can have a big house with a view of the reservoir. I'll take you there."

"You seem to know a lot," I said.

"Some, I missed." Barbara lit another cigarette. "After the coup, my mother sent me to Germany, nine years. I was seventeen. But my father stayed here. He was an agronomist. He helped the company run the farm that fed the workers. They had another German too, a professor who had been to Australia. This guy brought the eucalyptus to Brazil."

German parents, that explained the Brünhilde look. The rest of it was too much for me. What did eucalyptus have to do with anything? "Then you came back."

"I came back. Just before the first elections. Good timing."

"And you went to work for the city?" There seemed to be a gap.

"Oh no. I raised three children first. My father left his land to me. A hundred twenty hectares, you can see it along BR-040. For ten years, we had an organic farm there. *Then* I went to work for the city."

"But you're from here, right?"

Barbara laughed. "Nobody is from here. Três Marias is younger than I am. My family is from the south, from Santa Catarina. Where all the Germans are."

It was nice up here. You could see the reservoir and the dam and the town of Três Marias lower down, a few kilometres away. I wondered what the company would eventually do with the site but decided to leave that question for later. I was confused enough already. The woman standing next to me lived in a tiny house, had no car, eked out a living at a metal desk in city hall, but she had managed a three-hundred-acre spread — which she still owned. There didn't appear to be any Mr. Johnsen. One thing at a time, I decided.

"Those trucks," I said. "On the bridge, all along BR-040. What are they carrying?"

Barbara made a face. "Eucalyptus. From Australia, remember? They create huge plantations of it, then burn the wood to make charcoal. I will show you."

Introduced trees, introduced people — everything in Brazil seemed to have come from somewhere else, except the land. "Please," I said. I struggled to keep the conversation on track. "Why charcoal?"

"For the *siderurgica.*" She searched for the word in English. "You have seen them from the road. Where they make the metals."

"Smelters," I said. What I had written down as refineries. Grim places, with fire-belching mouths and trucks parked on flattened piles of slag. Charcoal trucks, I realized now. "That's all they burn?"

"Legally, yes. Also they burn the native trees, from the *cerrado.* Both are bad."

"So instead of deforestation of the *cerrado* you have a monoculture of eucalyptus. Where?"

"Where the *cerrado* used to be," Barbara said patiently. I felt stupid again. But at least I understood something. The charcoal trucks must be responsible for the calamitous potholes in BR-040. Trucks, refineries, *cerrado,* potholes: they were linked after all.

"You know what Manuelzão said when the first eucalyptus plantations went in?" Manuelzão was a local Mineiro, legendary for preserving the traditions of living upon the *cerrado,* for respecting and protecting the river. He died a decade ago, and his cottage, now a museum that Barbara helped create, is only ten kilometres from Três Marias. None of this, of course, I knew then.

"No, what?"

"He said, What will we eat, charcoal?"

Carlão drove us back down to my hotel, leaving the vapourized town to its breezes and its immigrant ghosts. I would never have stumbled on it on my own, not in a hundred years of visiting Três Marias, and of all the people I ever met in my subsequent visits, Barbara was the only one to mention it. Down at the reservoir the sun was just setting and the light was garish, a flaming orange. I took Barbara's picture on a bench, looking out toward the dam, her face rapt. She might as well have been gazing across Lake Geneva, not a man-made monstrosity that had brought more than its share of problems, social as well as environmental, to this place. Barbara was complicated; things here were complicated. Suddenly I felt sorry to leave.

But I had to go, and when I got home the first thing I did was take my slides to be developed. A week or so later I spread them out in rows on the light table, bending over each image with a magnifier. Half of them I threw away — lousy compositions, midday sun blasting away all trace of shadow and subtlety, bad guesses at exposure. The tail end of one of those charcoal trucks passing me on the bridge over the São Francisco; maybe that was Barbara's shit grass off in the corner. Barbara herself made the cut though, glowing preternaturally in the outrageous reservoir sunset.

But I kept coming back to a different picture, clearing the little illuminated field of bad tourist shots until a single slide sat alone. The guy in the car by the fish market, and his two kids. Both of the kids had dirty faces, I now saw, and something else: one of them was sticking his tongue out. At me. Welcome to the Rio São Francisco.

São Francisco snapshot

If you arrive in Brazil at Rio de Janeiro or São Paulo, as most travellers do, you're on the base of an equilateral triangle formed by three big cities. Rio, everybody knows, is on the Atlantic because

Antônio Carlos Jobim needed a place to write "The Girl From Ipanema." But São Paulo is closer to the ocean than most people realize. A couple of hours by congested freeway gets Paulistas to the beachfront hotels in Guarujá, only half an hour past stinking, fire-belching Cubatão, an industrial city once awarded the status of "most polluted place in Brazil." Cubatão in early morning is a forest of girders and tanks and flaming smokestacks silhouetted against the hills, like immense birthday candles. Sir Richard Burton spent three years nearby as British consul in Santos, now a petrochemical port smothering under a permanent pillow of yellow smog. The first thing he did was rent a house up where the air was better, in São Paulo, installing his wife, Isabel, there and journeying down to Santos for his weekly stint in the office.

The third big Brazilian city, at the top of the triangle, is Belo Horizonte. The name means "beautiful horizon," and Belo is capital of the state of Minas Gerais, which means "general mines." So, a nice view of a lot of mines, which was my first impression of the state; more accurately, a nice view of a lot of mines and hydroelectric reservoirs. It takes an hour to fly there from São Paulo, plenty of time to count dams and shake your head at hills laid open like some kind of primitive cranial operation.

The first time I visited Belo Horizonte I found the city chaotic, streets snaking off in all directions, many of them paved in chunks of rock that make your tires moan. Even after I graduated to navigating myself through the city by car, I never found any reason to modify this impression: the city's three million people seem to get by without a single street sign. Pampulha Airport is smack in the middle of it all, easiest to locate by simply looking up: "Over there, a big one, I can see its undercarriage, we must be close!" Looking up at airplanes is in fact one of the cheapest forms of family entertainment in Belo; you just park your car between some runway lights, open up the hamper and lie back. I don't feel so bad about getting lost there, because even a Mineiro like Carlão, Barbara's driver with the hundreds of nieces, can get flummoxed. He took me into Belo once, with a carload of his relatives, and became so flustered he actually reversed out of an off-ramp into a roundabout seething with cars, trucks and motorcycles. I remember turning around in disbelief and resigning myself to die. But nobody else seemed to mind.

The São Francisco River starts about four hours by car from Belo, in the hilly country of Minas Gerais. Like all good birthplaces, the spot is protected by a national park (the Parque Nacional da Serra da Canastra). And like most national parks, this one has somebody busily butting at the boundaries to get at something valuable inside — in this case a Canadian company convinced that De Beers, the world's biggest diamond consortium, made a mistake when they sold off their holdings there. This is diamond country, as is so much of the upper São Francisco basin. Brazil was, after all, once the world's leading source of diamonds. Fortunes have been carved out of the earth, and no visitor describes it better than Richard Burton: "It is a fracas of Nature, a land of crisp *serras* stripped to the bones, prickly and bristling with peaky hills and fragments of pure rock separated by deep gashes and gorges."

The diamond workings in Minas occupy Burton for several chapters of his book, *Explorations of the Highlands of the Brazil,* as though he couldn't get the prospect of riches out of his mind. It was a rough, desperate business, he wrote, and much of the exploration was conducted in the rivers, where alluvial diamonds were there for anyone prepared to wade in and sift through tonnes of gravel and mud. He found the Brazilian diamond extraction from rivers primitive, with "no trace of crane, pulley or rail, no knowledge of that simplest contrivance, a tackle, the negro was the only implement." It was dangerous, too: "Men who bathed in the diamond rivers were flogged, and those found washing in them lost their hands," and "an exceptional diamond generally counts in the wild parts [for] at least one murder."

But sometimes the payoff is huge. A third of the way down the São Francisco River, the Abaeté River enters on the left, its clear waters running uneasily beside the São Francisco's until the next bend finally mixes the two. If you're in a boat and you paddle a hand in the two waters, the São Francisco is noticeably colder, precooled by the reservoir upstream. It takes about an hour to get there by boat from Três Marias, and it's a location often visited by fisheries scientists. The last time I was there I spotted a telemetry antenna lodged high in one of the trees on the bank like a kite come to earth, picking up the signals from tagged *surubim* catfish. Not far from this spot, in 1800, the fist-sized Braganza diamond was sifted from Abaeté gravel, and in 1999 it happened again: *garimpeiros* pulled out the 79-carat diamond dubbed the Pink Star of the Millennium. The *garimpeiros* are still busy on the Abaeté, so busy that in early 2006 a crackdown on illegal diamond mining was announced. Fifty operations were photographed from the air, pouring sediment into the São Francisco. Most of them disappeared overnight.

The São Francisco River was named by Portuguese explorers who, as Sir Richard Burton says, "went down the coast with the Romish calendar in hand." But the river was not always called after a European saint; its banks nurtured many indigenous tribes, including the Cataguas and the Cariris. Once it emerges from the rocks in Canastra and drops over a waterfall, the river starts its northward journey and ends up flowing through six states before finally veering east to the Atlantic. Crossing all those political boundaries makes it a "federal river," which has all kinds of ramifications for its management. (In Burton's time, there was a serious lobby for creating a new Brazilian state out of the valley of the São Francisco — every city, town and village was "resolved to be the capital.")

The São Francisco collects water from 168 tributaries, many of which enter in the hills of Minas Gerais; the tributaries farther down the river, in the semi-arid northeast, dry up when the rains stop. The farther north and east the river goes, the drier the landscape, so that, in effect, most of the water comes from one end, while the other is parched. Fourteen million people live in the São Francisco basin, itself an area bigger than France or Thailand. That's 10 per cent of all Brazilians, and it's a lot of voters, 25 per cent of them illiterate. Half of them are Mineiros, people from Minas Gerais, where most of the water comes from. You couldn't design a better arena for conflict over water.

And the São Francisco has a lot of water. Some unavoidable numbers: it's 2,700 kilometres from the diamond prospectors in Serra da Canastra to the out-of-work fishermen in Piaçabuçu where the river empties into the Atlantic. The river can discharge an average 2,850 cubic metres of water per second, even after its regulation by the chain of dams that starts in Três Marias. For comparison, the average annual flow in the Fraser River is around 3,500 cubic metres per second and can explode to three or four times that during spring snowmelt; the Thames muddles past London at less than a hundred. A whole lot of calculation and negotiation has gone into producing another figure, 360 cubic metres of water per second, which everyone seems to agree is the amount that humans can take out of the river for whatever purpose they see fit. Those 360 cubic metres represent lesson number two about the river: there's only so much water available for people, and if you think counting angels on the head of a pin is absurd, you should see how many ways 360 cubic metres can be divided up. The arithmetic is secondary. We'll revisit the numbers later.

So: a big river, no doubt about it. It is definitely one of those "I wonder what *that* is" sights from high up in an airplane. It's true that the river has lost much of its primal force to dams and irrigation projects over the past fifty years: a cubic metre here, a hundred there. When you fly over an undammed river like the Fraser I grew up with, the plume of silt spewed into the ocean is astonishing, a good three kilometres of brown refusing to mix with the green of the ocean. In a boat you can practically reach down and thrust your arm into the dividing line, it's that distinct. A river like the Fraser seems determined to stay a river even after it's reached the sea. The São Francisco is different. At Piaçabuçu, on the Atlantic, the flow is so diminished it's as though the river is staggering to the sea on its last legs. In fact, the traditional fisheries of the river mouth have dropped to almost nothing as the salt water has begun to work back into the lower reaches of the river.

"Big" is relative of course, and as a way of describing natural features of the land it can easily become meaningless. Waterfalls, for example. How do you compare them? Is a thread of water trickling off a five-hundred-metre cliff bigger than a kilometre-wide thunderer a quarter of the height? You can charter a plane to fly you past Angel Falls in the tabletop mountains of Venezuela, but the experience is nothing compared to standing on the thrumming catwalk fifty feet from the lip of the falls on the Iguaçu River, soaked by spray and wondering how so much aquatic greenery can survive on rocks that are continually being blasted by thousands of cubic metres of water each second. The Iguaçu falls aren't very high at all in comparison with Angel Falls, the highest in the world. But Angel Falls is just a thread of water, and it really only knocks you out when you imagine Jimmy Angel, the discoverer, crash-landing his plane and walking back to civilization. *That's* impressive.

But people insist on rankings, so here they are for the São Francisco. The river basin itself — the land it drains through its tributaries — is the second largest in the country, after the Amazon. That's big even by Brazilian standards (it's the longest river that flows entirely within the country), and big also in comparison to better-known rivers. Officially, the São Francisco is number nineteen in the world by length, meaning half the Amazon. It's about the same as the Yukon or the Rio Grande, longer than the Danube, the Orinoco, the Ganges, the Colorado. What's more, in a nation given to nicknames (think Brazilian soccer players), only one river has one — and that's the São Francisco. Brazilians call it Velho Chico, which translates clunkily into "Old Frank" but captures the way Brazilians feel about the river, a feeling they don't extend to any other. "Our Chico" they say, in songs, in literature, in laments and in everyday speech from its source to its mouth. Brazilians love this river.

There is plenty of writing about the São Francisco, most of it Brazilian. Apart from the steady flow of articles and polemics about the river and its present problems (Brazilian academics and NGOs must be the most prolific in the world in this genre), revered Brazilian authors like Euclides da Cunha and Guimarães Rosa have made the river a setting for stories and novels that have never been out of print.

In English, though, there isn't a lot written. Germaine Greer produced an odd, angry chapter for a book on river journeys that accompanied a series of BBC programs in the 1970s. It's full of wonderfully inappropriate imagery like "the great yellow river writhing in its bonds" (the São Francisco doesn't writhe, it chugs), and "eucalyptus planted so closely that from a distance they look like dull green Axminster" (I like that one better, but when she follows it up by saying there's no space to walk beneath the trees she might get an argument from the guys with chainsaws who fan out into the plantations every day to cut them down). Greer and the BBC crew travelled up the remaining navigable section of the river on the old paddle wheeler you can still see tethered to the bank in Pirapora, near Três Marias; the boat was fired up specially for the film crew and it still makes the trip on festival days.

Thirty years after Greer's dyspeptic account, the BBC funded another writer to travel the river; this time, the project was an Internet blog by the Brazilian BBC reporter Paulo Cabral. Besides the obvious evanescence of an online journal, this account is also different from Greer's in being an example of the sturdy genre of travel writing in which the writer follows in the footsteps of a previous explorer. Marco Polo, Captain Cook, Amundsen and Scott of the Antarctic — lots of explorers have generated great travel books hundreds of years after the original journeys. But . . . the São Francisco? What explorer of any stature ever took the time to travel this obscure river and write about it in English? Well, why else would the British Broadcasting Corporation back the peregrinations of a Brazilian reporter? The inescapable Sir Richard Burton, of course.

Actually the São Francisco has had its celebrators over the centuries in all the arts, not just writing. Music is probably the most popular medium, and many of the musicians of the São Francisco are amateurs. Once, at a gathering in Belo Horizonte, I saw a guitar materialize in the hands of the unlikeliest of musicians: a local politician whose views on human rights, as far as I could make out, seemed not to have changed since those of the settlers who made such short work of the Indians. But the man could sing, and he gave us songs of the São Francisco, some of them his own compositions. The people he'd been offending listened happily. There are thousands like him all along the river, in the few big cities and the innumerable dusty villages, all singing their own songs of the river or those that have been passed down through their families. I've met more than a few of them; some record their own CDs, some actually make a go of selling them. But money isn't really the point. The point is Velho Chico.

DAMS, EUCALYPTUS AND SHIT

When I first saw this Chico, from that bridge vibrating with the parade of charcoal trucks, it didn't look like much. How, after all, can you hope to be captured by a place that's twenty-seven-hundred-kilometres long and pretty much inaccessible by car except at a few points? The dam at Três Marias was impressive, but the São Francisco is plugged by another six dams downstream. So, for that matter, are most major Brazilian rivers apart from the Amazon. Nothing special there. All I could see was a muddy, stopped-up, polluted river and some tired people.

The River of National Unity has been a mess for centuries, the insults accumulating as the interior of Brazil was opened up, goods began to flow and people began to settle. In the second half of the nineteenth century, steam vessels, railways and electrical energy all touched the river. Now, as it faces its greatest affront, the pharaonic diversion scheme I imagined Mike Holmes ripping apart in disgust, the river is already gravely stressed.

Water withdrawal is one big reason for the stress. Even though, theoretically, there's plenty of water coming down from the hills in Minas Gerais, the amount that's officially taken by agriculture and industry is throwing ecological equilibria out of whack. Water is also withdrawn illegally, for small mining operations, for small wells, and, most commonly, by farmers who simply build their own reservoirs. For every official permit there are a hundred illegal withdrawals. That 360 cubic metres of water per second, the officially authorized total withdrawal of water from the São Francisco, may be something of a holy figure for both sides of the water diversion war, but it's definitely understating the reality.

Another problem is deforestation. The *mata ciliar*, the natural vegetation that protects the river's banks, is 90 per cent gone, replaced by soybean and eucalyptus. As Barbara told me that first day above Três Marias, eucalyptus is grown solely to be reduced to charcoal to feed the smelters of Minas Gerais — six thousand tonnes of it every year. And those eucalyptus plantations account for only 60 per cent of the charcoal used in the furnaces of Minas; the rest still comes from native *cerrado* vegetation gathered illegally in the river valley.

Policing the area and the fourteen hundred kilometres of roads used to transport the charcoal is impossible.

On a later trip I was finally able to hold Barbara to her promise to show me that eucalyptus. We drove to a plantation thirty kilometres outside Três Marias. From the main road, a recently harvested stand looks like a clearcut, the uniform logs cut into sections and strewn like matchsticks all the way to the distant edge of the cutblock. It's only a temporary moonscape; in a month or so the stumps sprout again, like infernal weeds, and they put on an incredible two feet every month. Their only enemy is ants, but there are sprays for that. I nipped down an access road and squatted with my camera by the edge of a cutblock, fiddling to focus on the jumble of slash and the distant mohawk of the uncut plantation. Two men in a Gerdau company truck materialized out of nowhere. I played the stupid tourist while they wrote down the plate number and demanded my name.

"Schubert," I said. "Franz Schubert." Back in the car, Barbara just shook her head and laughed.

"Not here," she said. "This way." We left the highway a little farther on, passing through a tiny village before dipping into a much narrower access road that was almost immediately overhung with the slender grey trunks of eucalyptus.

"Now you can see everything," said Barbara, sitting back. I wrenched the wheel to get out of the way of a pickup truck with two filthy guys in the back, a couple of chainsaws rattling at their feet. It was just like a logging road back home, except that the road was sand, not gravel, and the rented car yawed back and forth like a ship in a following sea. Our wheelspan didn't fit the grooves worn by bigger vehicles — *much* bigger vehicles, it seemed — and the oilpan scrunched repeatedly into sand. Sun flashing through the eucalyptus trunks dappled the road and hid the potholes. Every few kilometres, we would come to a fork and Barbara would say, "That way — I think." The sun was close to setting, and there were a lot of forks.

I began to wonder how long this trip would take. After forty-five minutes the rows of eucalyptus became hypnotic. Every view looked the same. The underbody nosed up over orange crests and down through washouts filled with orange mud. The wheels spun.

Eucalyptus plantation near Três Marias.

I began to consider killing Barbara because we would certainly never find our way out after dark.

"There," she suddenly said. *"Fornos."*

They sat in a clearing, a line of forty or so beehive-shaped clay ovens, each one the size of a couple of trucks. Stacks of eucalyptus logs lay around, ready to be fed into the ovens, and the smoke from all those roof holes joined together into a single yellowish cloud. From an airplane it would have looked as though the forest itself was burning; who knows how much carbon dioxide was shooting up into the atmosphere. There was nobody here, only a friendly black and white dog, but I could imagine the scene: the loggers with their chainsaws dumping the sections of eucalyptus, and the big trucks wallowing down the sand river to pick up the bags of charcoal. Now I knew why Barbara had brought me here late in the day: the charcoal trucks were gone. They had already become, for me, a menacing icon of the destruction of the *cerrado;* here I was now, in their lair. We got back in the car to try to find our way home.

"The owner lives in Santa Catarina," said Barbara. Warm sand, turquoise water, a long way from here.

Dams, water withdrawal and deforestation are bad enough, but the São Francisco also suffers from pollution. Industrial effluent enters the river in many places, both from tributaries and the main-stem banks. Downstream from Três Marias, in 2005, an outcry erupted over the corpses of catfish that had started surfacing, sus-pected victims of a non-functioning settling pond at the zinc refinery that even I couldn't miss on my first visit to the town.

The problem doesn't stop with industry. Only 54 per cent of Brazilians have sewage treatment, which means that more than four hundred municipalities in the basin simply dump their untreated waste into tributaries. It's harder to point the finger here, because dumping raw shit is a common practice in many countries. It's not as though using rivers for sewage disposal is something that's only done in shabby places that can't afford anything better or don't care enough about the consequences. Having flowing water nearby is a powerful invitation to dump stuff into it; my own country does it all the time. Elegant Montreal flushes its sewage into the St. Lawrence with only primary treatment. Sophisticated Vancouver uses the Fraser River and the sea, with only partial treatment. And Victoria, where I live, still pumps sewage a kilometre into the sea and lets the currents and the micro-organisms do the rest. I remember walking the beach as a kid and picking up a funny wrinkled thing with an intriguing ring at one end. "Mom," I asked, holding up my latest find, "what's this?" I don't recall her answer, but I do remember her jumping backward. These days, we screen out the condoms, and there are even plans to stop using the ocean altogether.

WRITING A TANGLED TALE

How many insults can a river like the São Francisco withstand before it's just a channel between two banks, an aquatic supply line to plug into? What's happening there now is the kind of horrifying, slow-motion train wreck that results when science, politics and greed climb into the same bed. The story of the battles over the São Francisco is folly on a grand scale. If the river were a character in a play it would be star-crossed, marked for greatness and inevitable disaster, a lesson to humankind if humankind ever heeded such things. Maybe that's why, after my first few trips there and the beginning of my re-education at Barbara Johnsen's hands, I began to dream about writing it.

But to tell such a story means writing about a lot of other things: history and geography and, especially for me, science. Fisheries too, because you can't describe a river without describing the things that come out of it. And because river fisheries make the most sense in comparison with the fisheries people know best, the ones in the ocean, that means writing about oceans too, and waters in general, both fresh and salt, and the way they're tied together in immutable global cycles that rule all of our lives. It's a big, tangled, incestuous mess of a story, and if you look under enough rocks you'll catch a whiff of inbreeding: everything is related to everything else.

The São Francisco runs through six states, and the whole country is watching it. But the agony extends beyond Brazil. What's happening to the São Francisco now is a drama that's being staged in plenty of other places around the world, even if the negligible profile of rivers means few people are aware of what's going on. But because the São Francisco's problems are the problems of the world's rivers, to understand them means meeting bit players from around the world: the expatriate Brits and Norwegians, the development bureaucrats we've already seen marking their territories in Rome, the experts I call science tourists. The truth is that a hatchery manager in the interior of British Columbia and a guy in boots and bloody apron slicing a six-foot tuna with a sword in Tokyo have as much to do with fisheries in the São Francisco as do the Ministry of Agriculture bureaucrats in Brasilia. People, fisheries, rivers, oceans, water: they're all connected.

Maybe that's why it took me so long to dive in. But the São Francisco is where I kept returning, and it flows through this book. How I ended up there is part of the story, and I'm certainly not the first to have been drawn to the place. Sir Richard Burton will raft through this book too, bobbing up and down, notebook in hand. His journey, and the book that resulted, are like a bittersweet chorus: he saw the past, he recorded the present *(his* present) and, despite himself, foretold a sorry future.

As for the plot, it has always been Byzantine, what with the river's central place in the last three centuries of the country's history; and that plot has thickened over the last few years with the rebirth of the mother of Brazilian megaprojects, a multibillion-dollar network of canals and pumps and aqueducts to tap the upstream waters and redistribute them in the thirsty downstream northeast. Do the benefits of damming and diversion outweigh the elimination of ecosystem functions? It's pretty difficult to argue that zero development is the only way, not with the demands we all make for cheap power and produce.

The real question is, How much is enough? When does it become too much? As usual, Sir Richard has something to say, although he is characteristically hard to pin down. On the one hand, he was a man of his time, the ex-agent of the East India Company looking out for opportunities. But he was also a loner, happiest with his notebook and the awesome silence of nature.

"They are emphatically the Lands of Promise," he writes in 1867 about the valley of the São Francisco, upon retiring on a sandbar for the night. You can almost see him reclining on one elbow, listening to the river slide by, writing by the light of an oil lamp. "This desert stream will presently become a highway of nations, an artery supplying the lifeblood of commerce to the world. The 'Lion' Rapid and the 'Fierce' Sandbar will be silenced forever. And the busy hum of man will deaden the only sounds which now fall upon our ears, the baying of the *guará* wolf, and the tiny bark of the little brown bush rabbit."

What a poignant mixture of the love of wildness and the vision of an age! The baying of the wolf had certainly been silenced by the time I first visited the river, and it would take several more visits before I could begin to see the river through Burton's eyes. Brazil was a big place; there were many things to distract me from a dying river. I had some meandering of my own to do.

CHAPTER 4

WATERS

Oceans and rivers are apples and oranges

A pleasant place to be, unless you're a fish: local haul on a beach in Santa Catarina, Brazil.

"SALTO!"

That's what it sounded like, anyway. But *salto* meant "a jump." Maybe it was *salte*, which would have made it the imperative form of the verb *saltar*, which meant the guy wanted *me* to jump. Lucia Lopes's conversation classes hadn't covered any of this territory, but "jump" made more sense, and it explained the gun. The guy said it again, and this time I caught the "a" at the beginning: *assalto*. That made even more sense, and it explained why he sounded so impatient. *Assalto*. Robbery. This was a stickup.

I haven't seen that many real guns. As far as I know, this was the first to be pointed at me. It looked awfully big, dull silver and filling the space between my body and his. It looked heavy, too; maybe that's why the barrel was floating around between my chest and my knees. I let out a humiliating little fart of pure fear, just like in the books; I couldn't help it. There were just the three of us: me, him, my wife, Hatsumi, on her first trip to Brazil. All alone in the middle of a beach with the sun going down. A nice place for a stroll, if you kept your eyes open.

But I hadn't. Worse, if I'd been looking for a place to absolutely guarantee getting robbed, I couldn't have come up with a better one. A long expanse of sand, on one side a bit of scrub with the occasional small hotel and footpaths to the main road. On the other, the Atlantic. Plenty of people during the day, but the light was failing now and the place was deserted. Any idiot would have seen the gunman a mile away — literally. And here he was, skinny and shuffling from one flip-flop to the other and waving a gun in the general area of my navel. Maybe he'd just appeared, beamed down by some exasperated higher power to drive home a point before we left the next day. Stranger things have happened in Brazil.

Ilheus, where we were, was a break from the São Francisco River. Ilheus was on the ocean and that meant it was a vacation, a place to go back home from. Two hours earlier, this same stretch of beach had been alive. It was Saturday, and it seemed the whole city of Ilheus and whatever tourists came here were recreating furiously. No *gringos* either; whatever allure Ilheus may have by virtue of being the birthplace of Jorge Amado and the setting for all his stories, it wasn't pulling the Germans and Americans away from Salvador da Bahia and Rio. Young Brazilian guys were playing the inevitable game of pickup soccer on the sand that afternoon, the goal posts a pair of sticks only a metre apart yet still bisected by shots drilled barefoot from ten metres. I watched an out-of-bounds ball chased toward the sea and, instead of being corralled and brought back, thumped high over the surf for the sheer joy of seeing it launch toward the sun. Beneath the arc of the ball, a grinning little girl in a bathing suit rode a grown-up bicycle, standing up but still barely reaching the handlebars. We heard yells, shrieks and nodes of thumping music around the cabanas serving food and beer, the whole beach in constant motion as groups of people ambled from one end to the other. One man dropped suddenly to the sand and threw himself into a series of sit-ups, as though trying to eliminate a few Antarcticas' worth of calories.

As the sun began to settle, the spent waves flooded the sand like poured metal, and ghostly pale crabs seemed to float over the surface, visible only by their movement. A black cloud condensed over the sea, then a sudden, chilling wind drove people off the beach into a *cabana*. The rain arrived horizontally. Waiters jumped to block it with beach umbrellas deployed around the tables; an old leathery man in a codpiece bathing suit nearly hidden under his paunch danced until the storm passed. In the middle of it all, we ate Bahian *carranquejo* served in ceramic crabshells. And then, stupidly, we decided to go for a walk.

The robber looked about sixteen. Skinny, in a white T-shirt and baggy red shorts. I looked once in his face and didn't try it again. Dark skin, short, straight black hair, a gap between his two front teeth, nothing special except the eyes. They were dead and unfocused, like a doll's. I dug in the right pocket of my shorts for the sacrificial wad of reais I always carried. Maybe twenty dollars. He took the money and waved the gun at my other pocket. *"Assalto,"* he said again, in case I hadn't got the point. His voice was dead too.

Finally I started to get angry. The other pocket had my Petzl headlamp, the one I took everywhere for reading in bed in strange rooms and getting up to pee and, great international traveller that I was, walking back to my hotel at night on deserted beaches. I loved my headlamp. I pulled it out and dangled it by its strap.

"Só um luz," I said, it's only a light, using the wrong word. I should have said *lanterna.* But it didn't matter, he took it anyway, stuffing it in his shorts. Probably he would chuck it in the bushes. He tucked the gun under his shirt and angled off across the sand, toward the road. He didn't even bother to run. We would have to remember our way back, find the hotel's footpath through the scrub by the light of the moon.

"Shit," I said to Hatsumi. "He's got a friend." We watched the second shape detach from the bushes between the beach and the road, merge with the robber's silhouette advancing across the sand, stop for a moment and then vanish, probably back to the city a few kilometres down the road. No point in working the beach any more tonight, we were so obviously the last ones out.

"Sorry," I said. "Sorry, sorry, sorry. I should have seen him coming. Did *you* see him coming?" Hatsumi shook her head and took my arm. We looked around. The way we had come looked a lot blacker and emptier than it had five minutes ago. The rollers coming in all the way from Africa were starting to pick up a little moonlight. "I vote we try the road this time," I said, and we headed for whatever pathway the robber had taken. Five minutes later we were walking back the way we'd come by car the day before, looking for the perfect *pousada* to spend a day off before flying back to Canada. Green buses packed with people leaving the city rumbled past, windows open. I looked at my watch: even this robber hadn't been interested in my shabby Timex. Seven-thirty, the night was young. Dried-out palm fronds crackled under our feet.

When we finally saw the illuminated sign for our *pousada,* we went straight through to the back and sat outside on the small patio that looked out to the ocean. The landlady brought beers and went into the kitchen to make dinner. The wall beside our table was studded with shards of broken glass embedded in the masonry, like all the properties around here.

"Tell you what," I said. "I'll call Arley. At least we should let somebody know. Maybe there's something he can suggest." I got up and climbed the stairs to our room to fetch the rented cell phone. Arley Ferreira was a captain in the *Polícia Militar* in Belo Horizonte. He'd seen us off at the airport; his sister-in-law had helped us find this *pousada.* Arley was responsible for environmental policing on one stretch of the São Francisco River. He knew we were here, and he was a worrier. If I didn't call, eventually he would. He answered on the first ring.

"O Arley." I love the formal mode of address in Brazil: The Arley. When I first came to the country, I wondered why everyone thought I was Irish.

"O Brian, *tudo bem?"* How's everything? I went right into it, leaving out the part about his sister-in-law leading us to this dangerous place. I asked him if he thought the gun was real. "Real?" he asked. Did I think the robber was playing? "Now you must go to the police station and file a BO. For the *cidadania."*

I could imagine Arley at home, brow furrowed under close-cut grey hair, the perfectly pressed tan uniform replaced by shorts and T-shirt that were probably perfectly pressed too. A BO was a *Boletim de Ocorrência,* a statement that something had happened. An arrest was unlikely, but it was a citizen's duty — *cidadania* — to line up for three hours and fill one out. *Cidadania* meant a lot to Arley, as it does to most Brazilians. It makes a weird kind of sense in a country of so much pain, such inequalities, such a deeply entrenched oligarchy of unelected rulers descended from the first Europeans to grab the best land from an inexhaustible supply. If you didn't believe in something like *cidadania,* or even the hope of it, you would go mad. Brazil is also the land of citizen-heroes who have fought — and continue to fight — a succession of oppressors, and whose victories and martyrdoms are the stuff of stories and songs every Brazilian knows. Arley clung to the idea of *cidadania* the way I still clung to the idea of science; somehow, if we believed in them enough, they would see us through.

"Okay, Arley," I said. "I'll do a BO." I went downstairs and told the landlady instead; it was probably as effective as telling the police. "It's the first time for me," I said. This wasn't strictly true, if you counted the time a taxi driver had my gear under his car at the Caracas airport before I had the wit to get out and demand everything back. But I'd been awfully lucky over nearly twenty years of travelling, a fact I'd always attributed to my superior watchfulness. I looked at Hatsumi staring out over the black Atlantic. I could almost see the thought balloon over her head: "He told me it would be safe here . . ."

But the *muqueca* was delicious, a Brazilian bouillabaisse with coconut milk. The ocean thundered reassuringly in the dark, and a lone ten-year-old sat in the shallow end of the small pool behind us running an open hand back and forth over the underwater light so that the blue rectangle flashed like a jewel in its setting of tiles. The beer was good and it arrived in very large bottles.

The next day was Sunday, going-home day, a day that would end over a cold Canadian runway glimpsed through sideways streaks of January drizzle and me wondering, as always, whether I had really got up in the impossibly distant tropics. When it was time to leave for the airport the streets had begun to fill but there were no taxis, so we trudged to the bus depot on the edge of town, a strip of wood and canvas shelters with soft drinks and snacks and, after patient interrogation, tickets. The municipal buses seemed to go everywhere, even as far as the beach at Olivença twenty-five kilometres north. They all cost fifty cents. At the airport, a girl selling coconut milk from a green cart kept her eyes on the soap opera blaring from an overhead screen. There was no security check.

We hopped along the coast to change planes in Salvador, then angled inland over cane fields south to São Paulo, descending through a yellowish haze over a forest of orange-tiled roofs that yielded to a snaggle-toothed carpet of concrete. In São Paulo there would be a long, sapping interregnum; the flight to Canada leaves near midnight. As usual I looked down, for rivers. The story I was already beginning to write in my head was about water, fresh water, the kind that seeps out of the ground unnoticed and collects imperceptibly in rivulets and springs and tributaries and finally in the staggering volumes we call rivers, pouring downhill any way it can, slaking thirst and feeding crops and picking up crap and chemicals before it reaches the sea. Strange for someone who grew up with Captain Jacques Cousteau as the shining example of all that was romantic about field biology, but somehow, over the years, I'd stopped working in the marine environment and found myself spending my time where the water flowed downhill.

But our approach to São Paulo this time was from the ocean, over Santos, where Sir Richard Burton spent his term as British consul in the 1860s. The only river I saw was the miserable Tietê, strangled and canalized by the city. That was the problem, I thought. Rivers just don't get any respect.

POPULARITY CONTEST

When I tell people I work in Brazil, the reply is always: "Wow. The Amazon. See any piranhas?" This is usually followed by envious sounds about Rio and beaches. If they find out I'm in fisheries they also want to hear about the *candirú,* the little whatever–it–is that swims up your penis. For most people, Brazil seems to be a burning rain forest surrounded by a sandy beach, and when I try to describe the São Francisco and the interior of the northeast I usually lose my audience. The Amazon always excluded, what is it about rivers that makes them so unromantic?

"Roll on, thou deep and dark blue . . . river?" It doesn't sound right, somehow. "Ten thousand fleets sweep over thee in vain," Byron wrote next in *Childe Harold,* which is hard to imagine on a river unless the boats formed a conga line, or were remarkably small, or it was a freshwater lake like the Tonle Sap on the Mekong, sprinkled with fishing canoes, in which case Byron could actually foretell the future. To be fair, another line from *Childe Harold* — "Man marks the earth with ruin but his control stops with the shore" — could pretty much apply to rivers as well as oceans if it were true anymore; so, too, with "Time writes no wrinkle on thine azure brow, Such as Creation's dawn beheld, thou rollest now." But both were written about the ocean.

Did Paul Gauguin hole up by some muddy South American river to paint his bougainvillea and his child-brides? No, he went to Tahiti. Was Malcolm Lowry to do his best writing on a gravel bar on the upper reaches of the Nahanni? Of course not; he built his shack on the beach around the corner from Vancouver, where he could rinse away the night's whisky with morning plunges into the icy water, then row across the inlet for more. Here is Melville, describing the Pacific in *Moby-Dick:* "The same waves wash the moles of the new-built Californian towns, but yesterday planted by the recentest race of men, and lave the faded but still gorgeous skirts of Asiatic lands, older than Abraham; while all between float milky-ways of coral isles, and low-lying, endless, unknown Archipelagoes, and impenetrable Japans."

Poets normally reserve this kind of magnificent language for the oceans, but even popular expressions tend toward the marine. We have "oceans" of fun and if we're really blissed out we get that "oceanic" feeling, but what happens when we lose it all? Then we're drowning in a "river" of debts. "Cry me a river" is typically negative, although Brazilians use the metaphor of tears a little differently. "Our river is crying," they say along the São Francisco, in newspaper articles, in songs, in everyday speech. "Cry me a river" makes me think of how rivers start, inconsequentially like the *drip drip drip* of meltwater off the dirty tongue of an expiring glacier, or the leaking earth that is the start of the São Francisco in the Serra da Canastra. A river is like a billion billion tears, and where do those tears have to go before they're worthy of a decent quote? To the ocean, of course! The image goes to the heart of one of the big differences between rivers and the sea. Rivers start somewhere and they end somewhere and they can jump their banks and ruin you, but the ocean is limitless.

Rich people swim in the ocean, but poor people swim in the river — along with whatever filth is arriving from their neighbours upstream. In Três Marias, houses shoot their sewage into the foul little creek that connects directly with the São Francisco; one time

I saw a dead cow and a dog or two circling slowly in the water hyacinth at the mouth, as though waiting for someone to call them home. In the next five kilometres, the river receives effluent from a zinc refinery and runoff from plantations of soybeans, melons, grapes, eucalyptus. The capacity of the oceans to accept waste, while not infinite, is still infinitely greater than that of rivers.

Rich people eat marine species like the crab we enjoyed in Ilheus; poor people eat river fish. When the average person thinks of seafood, they think of lobster or seared ahi tuna, not muddy-tasting catfish or a stew of oily, fruit-eating *pacú*. People certainly don't fish on rivers for the view either, at least not on tropical ones where the liveliest thing may be a clump of vegetation drifting downstream: look, here comes another waterlogged branch. It's true that rivers are geographically impressive and they cover a lot of territory; the water rushing past you on the bank may have travelled farther than the wave breaking on a sheltered beach. But that water is usually turbulent and often downright opaque, so even if there are brightly coloured fish somewhere in there (and there are — ask any hobbyist where their cardinal tetras come from), you're unlikely to see them even if you were crazy enough to put on mask and snorkel and dive in.

Rivers are confined, the antithesis of the dizzying expanse of the ocean where navigators have to remember that the surface of the earth actually curves, that the buoy or landmark they're straining to see may be hidden simply by the ocean's size. Trapped between banks, rivers become connectors, highways, magnets for colonizers and, of course, sources of water to drink and to pour on your crops. This is what happened in Brazil. When the Portuguese landed at Salvador da Bahia in 1532, not far from Ilheus where I took my ill-fated vacation, they soon left the gorgeous vacancy of the beach and set off inland, killing Indians, building churches and miscegenating busily on the orders of their king. They left the ocean behind and threw in their lot with a hot, confined, unpredictably flooding channel that became, four hundred years later, the River of National Unity. For them, the ocean represented escape. The river was reality.

So we have a planet covered in water, most of which is ocean. People catch and eat things from both kinds of waters, oceanic and

Rivers are more impressive when they're vertical: Iguaçu Falls, Brazil.

inland, and in the last decade or so the problems of some of these fisheries have been spectacular enough to earn media attention. Most of the "big stories," like the collapse of the northern cod and the winking out of salmon runs throughout their global range, have been in the developed world, although the eradication of three hundred species of native perch in Lake Victoria was also international news, perhaps because it resulted from a misguided attempt at "development." In general, though, attention stayed on ocean fishing, mostly in developed countries.

It's hard to get worked up about fish of *any* kind because you can't see them — out of sight, out of mind — and they're the opposite of cuddly. A conservationist concerned about aquatic species has to make some hard choices. With the exception of whales, conservation of animals has been mostly about terrestrial mammals and birds; pandas and snow leopards get the ink. Of the aquatic species, marine ones get the most attention because they're colourful and the ocean's an attractive place — something Captain Cousteau knew very well. Marine creatures are more photogenic than fresh-water ones, and it's a lot easier to put a camera on them. Documentaries about the oceans have become pure nature porn, with endless sequences of dolphins arrowing into the sunlight, manta rays shot from below so that they create an ominous outline against the surface, shoals of panicked sardines collapsing and reforming by the split second, like a gigantic superorganism. Against this, what do tropical rivers have to offer? Only the water in them.

Marine species are also easier, and usually a lot more pleasant, to study scientifically. For example, wonderful things can be done by implanting micro-transmitters in fish. Tracking systems for high-value species like salmon are proliferating everywhere the fish are harvested. Sea turtles are forever being interrupted in their egg-laying (on lovely white tropical beaches, by the way) and jabbed with satellite tags; and the latest research has just revealed that a single satellite-tagged bluefin tuna actually crossed the Pacific three times in less than two years, showing that "Japanese stocks" and "American stocks" are actually one and the same.

Against this, the effort to tag the movements of migratory river fish is barely out of the gate. Not because those movements aren't important (they are, to the fish and to people) and not because there's some technical hold-up. The procedure is the same whether you're on the ocean or in a river: catch a fish, inject a chip or cut a hole in the gut and slide in a lozenge the size of a fingertip that con-tains a microchip and battery and trails a six-inch antenna out the side of the fish like a wayward mooring line. (In Brazil they use clove oil to anesthetize the fish, so that if you closed your eyes you would swear you were in a dentist's office.) It's just that river fish are so ... unglamorous. If you had to measure the migratory movements of a fish — any fish — which would *you* choose? Something that swam in the ocean (think: exotic ports, salt spray and suntans) or something from a river (think: waterborne diseases, sunstroke and boiled catfish for dinner). There is no contest.

BUT FRESHWATER FISH ARE SEXY TOO

So oceans are sexy and rivers aren't. But here is a problem, a big, counterintuitive one: of all the aquatic species on Earth, nearly half live in fresh waters. Those shimmering iridescent shoals of amber-jack parting for the camera on a coral reef? Those photogenic fleets of blunt-nosed beluga whales gathering in the Arctic like a pack of white submarines? Forget it: drop for drop, the richness of aquatic life is actually greater in rivers and lakes. And because it's all happening in less than 0.5 per cent of the volume of the oceans, that makes freshwater habitats astonishingly precious. It's not surprising that, of the aquatic animals that make the Red List maintained by the International Union for the Conservation of Nature (IUCN), freshwater species outnumber marine.

What goes on in the muddy waters of a big tropical river is hard to study and drearily unsexy. Unravelling a typical food web, for example — figuring out who eats who — represents a decade of sampling in small boats in the hot sun or pouring rain. It means bottling and labelling and identifying, dissecting stomach contents and drawing lines between every single creature, one after the other; connecting zooplankton with shrimp, and mussels with eels until you have a constellation of maybe fifty species sewn together by a forest of arrows and pathways that looks like a circuit diagram without a right angle in it. It's just excruciatingly hard to learn anything about these remote systems, but it's absolutely necessary as long as we keep monkeying with aquatic ecosystems in the name of fisheries and development. Harvesting a few species from a complex web like this is like reaching into a tangle of phone cables and yanking out a handful of wires: what happens is hard to predict, and you need to go through the same laborious process to measure it.

Is there *anything* glamorous and exotic about freshwater fishes? As a matter of fact, there's plenty. The migratory species that are such an important part of life on the São Francisco are long-range cruisers with an infinity of shapes and sizes and habits, feeling their way along the muddy bottom or arrowing up through rapids and snapping at falling fruit. In the aquatic world, a six-foot *surubim* catfish takes a back seat to no one. And if you really must go outside the

São Francisco, to the Amazon, there's always everybody's favourite, the *candirú*.

Most people, when going near fresh water in Brazil, worry first about the big reptiles. Caiman, like armour-plated submarines cruising at periscope depth, you see them everywhere in the Pantanal, even crossing the road at night, when their eyes light like a cat's. If they're small enough you can get right into the river with them, although all you have to go on to judge their size is two eyes breaking the surface like wet grapes. Anaconda are another worry: eight metres of brainless malevolence looking for a leg to fasten onto. If you have a sharp knife and a lot of presence of mind, you can defeat an anaconda by waiting until it has eaten its way to your knees, then slitting the hinge in the jaw. So I'm told.

But there are a half-dozen fish that are just as dangerous as the reptiles. Piranha, everybody knows about. There are more than a dozen species and each one is plentiful in Brazil. Burton calls it the "dreadful piranha," adding that the name means "scissors" in the Tupi language and that he found the meat "dry, full of spines, and with poor flavour." One of the best ways to provoke a piranha is to corner it in a net. A net full of trapped piranha is like a sackful of razors; you can see — actually *hear* — their jaws working at the mesh. I once helped pull a net through a shallow lagoon off the Orinoco in Venezuela, and the thing writhed with piranha. The fishermen didn't take chances; they whacked them through the net with lengths of wood, one by one, like batting practice, really putting their backs into it as though paying back for old bites on their legs. A lot of the men had scars on their calves where a cornered piranha had struck. The chunk of flesh they tear out is almond-shaped, like a lozenge. I remember wishing I had worn something other than sandals.

There is more than piranha to worry about. Electric eels, as long as a leg and lurking in the river mud, can fire a bolt powerful enough to kill a child. Steve Irwin, an Australian better known as the Crocodile Hunter, lost his life to a stingray on the Great Barrier Reef, but there are freshwater rays too. Mud-brown and perfectly camouflaged, they're invisible until you step on one, then they lash backward with tails like buggy-whips, driving a barbed stinger firmly into your Achilles tendon.

But piranha, electric eels, even rays are tangible foes. You can learn where they hide. You can beat the water with sticks to scare them off, or stay away from nets. The *candirú* is different. A *candirú* doesn't play by the rules. Most people think it's a worm, but it's a fish. An adult is about five centimetres long but only a few millimetres wide; its body is slimy and scaleless and instead of jaws it has a round sucking mouth and rasping teeth like a cat's tongue. *Candirú* are parasites, bloodsuckers that feed directly off the gills of other fishes. They are also attracted to the warmth and taste of animal urine.

Since the earliest days of South American exploration, reports circulated of a diabolical fish that would swim up the urine stream of a bather foolish enough to answer the call of nature while submerged. The fish would keep going, striking deep into the urinary canal and heading for the bladder. The unlucky person was said to experience, at first, no more than a tickling, which increased to discomfort and then to pain as the parasite began to feed. Upon trying to pull the offending creature back out, pain became agony. The only cure (for men, anyway) was amputation. Natives, it was said, took to wearing codpieces made from coconut shells.

By the 1940s, scientists had identified the *candirú* as *Vandellia cirrhosa,* a member of the endlessly inventive catfish family. They found that the appalling pain came from the fish's tactic of expanding its gills when tugged, so that it became perfectly barbed and impossible to dislodge. *Candirú* seemed to be attracted to all human bodily secretions, with a special taste for urine, and they attacked livestock as well as men and women. The *candirú* is plentiful in the Amazon and Orinoco, which makes it common in a quarter of Brazil. My friend Hugo told me that if you leave a dead fish hanging over the side it'll be bled dry by *candirú.*

"You can pick them off and hang them from your fingers like Christmas tree ornaments," he said. "And if you just hold your hand in the water they try to swim up the cracks between your fingers." I've never seen one and they're not supposed to occur in the São Francisco, but one of the mantras of conservation is that "new species are being discovered every day." I'll continue to pee outside the river, just in case.

Don't mess with my ecosystem

How important are fish and aquatic ecosystems? It depends on who's asking — people or fish. When you look at global food consumption, around a billion people get their animal protein from fish, mostly in the developing world. In the developed world, inland commercial fisheries have largely been replaced by recreational ones, but they are still going strong farther south — even increasing. Unfortunately, some of these fisheries are out on a very long limb. Fishing a species into an evolutionary black hole, as is happening now, does a lot more damage than putting fishermen out of business. First, if it's a top predator that gets beaten down, that throws the rest of the ecosystem out of line: prey can explode, out-competing other fish lower down in the system, with the fallout extending all the way down to the algae so that primary productivity, the engine that drives everything in a water body, is out of balance. There's a new vocabulary for this kind of ecosystem effect. It's called "fishing down the food chain" or, even more ominously, "trophic mining."

One food chain pulls in another in southern Brazil.

Both of these wonderful phrases describe the cascading effect of harvesting a single species beyond its ability to replenish itself. If you look at global fisheries in terms of the trophic level of what's being caught (that is, where it sits in the food chain), it becomes clear that we are catching fewer of the big, fish-eating species and more of the small, plankton-eating ones. It's a steady decline, not a dramatic one, and easily overlooked, but it's been happening in marine as well as inland fisheries. Worse still, industry has simply fished harder and deeper, going after species that have so far been left in peace and that can sustain only the briefest of booms before the big nets have to be dragged somewhere else. Either way, aquatic ecosystems are being irreparably damaged. So fish matter not just to people, but to other fish.

Fish don't even have to be alive to serve as nutrient conduits. The salmon that die in rivers after spawning are actually packets of minerals assembled thousands of kilometres away, in the ocean, and these packets get spread around as fertilizer in lakes and rivers. The dead salmon also feed many other animals: aquatic ones (mainly small invertebrates) and terrestrial ones including bears, wolves and eagles; when their carcasses are dragged into the trees they fertilize those too. The massive migrations of some South American fishes, up and down the river and from main stem to marginal wetland, manage to connect ecosystems that are thousands of kilometres apart. A lot of these species don't even have to move in order to make the linkages between aquatic and terrestrial ecosystems, because they eat the fruit that falls from trees. The seeds wash up somewhere else, so that if you cut down the majestic fruit trees that line the banks, as has been done along so much of the São Francisco, you also sign the death warrant of the *pacú*.

Sometimes the consequences of fish collapses are bizarre. In 2000, the sockeye salmon failed to return to Rivers Inlet in northern British Columbia, and what happened next went far beyond people losing income from a catch that never materialized. The grizzlies missed out on their annual feed of spawning salmon, something they depend on to fatten up for the winter. So they did the most sensible thing: they shopped somewhere else. Unfortunately, that meant heading into populated areas, and when the bears began showing up in people's gardens, many of them had to be shot. Fewer salmon, fewer grizzlies.

Such situations aren't easy to fix without considering the ecosystem as a whole, and fixes can backfire. In Glacier National Park, when lake production of sockeye salmon dropped, shrimp were added to the lake to help feed the young salmon. This seemed like a reasonable solution: they're hungry, let's feed them. But it didn't work out that way. The shrimp beat the young salmon to the available zooplankton, so even fewer salmon were produced. The net result was a lake full of freshwater shrimp, and the lack of salmon meant no bald eagles, and fewer coyote, mink, deer and grizzly. A lake that generated tourist revenues now had no anglers and no bird-watchers, because the attempts to rebuild a single species didn't take into account the interconnections between all the others.

SHIT HAPPENS TO RIVERS

River boundaries change naturally with the season, so there's a constantly shifting relationship between creatures of the water and the land that surrounds them. And the land does surround them — that's why they're called inland waters. Rivers are just two long coastlines carved out through dry land, and this means that river species are essentially captive, vulnerable to whatever humankind wants to do to the habitat that confines them. River life suffers because people cut down trees, build factories, dump agricultural and urban waste, pull out the water to irrigate their crops and run their turbines. In the ocean there's a significant dilution effect: the sewage from the city of Victoria gets flushed far out into the formidable currents of the Strait of Georgia where it's debatable if there are any bad effects beyond the general feeling of embarrassment that we're blending our turds into soup and pouring it into the sea. In the rivers, there isn't much of a dilution effect at all. Your shit still flows downstream, and in a very restricted amount of water. The end result is that, species for species, the freshwater animals are more at risk than the ones in the sea.

Inland waters bear the brunt of the natural tendency of people

Living in the Rio Negro, Amazonas.

to colonize from the coast inward. Ocean coastlines are actually quite limited, a good thing for real estate agents but bad for human expansion. Societies develop by pushing inland, away from the ocean, laying down lines of communication and travel so that goods can flow back to the coasts and trade can flourish. So river basins accumulate people and their water needs — for power, for agriculture, for convenient waste disposal and transport. Colonization of Brazil by the Portuguese was absolutely typical: they arrived on the coast, at Salvador da Bahia, established a foothold and fanned out into the interior, following the São Francisco and its arteries back toward the source thousands of kilometres away. Water was always the key: too far away from it and settlement was impossible.

Inland waters are at risk because the corridors they flow through have high economic value for agriculture, for mining, for timber harvest and for industry. Many of these activities have multiple effects. Forestry, for example, removes ground cover, increases siltation, raises water temperature by eliminating shade, adds pollutants. In many cases, removing streamside trees actually removes food sources for fish (like the migratory *pacú* that live off fruit that falls into the river) or creates barriers to migration (like the choked streams that stop salmon before they can reach spawning grounds). And, most obvious and painful in terms of sheer human suffering, deforestation is a prescription for floods. This is the most gruesome of environmental Catch-22s, because the remedy for floods is, of course, dams.

Rivers feed cattle and soybeans and factories and people; they're magnets for civilization. They're political too: after thousands of years of shifting boundaries, rivers have ended up not only heavily colonized but also broken up into a patchwork of jurisdictions. Almost three hundred of them cross international boundaries. Supplies get squeezed, pressure points emerge, the value of fresh water goes up. The infrastructure for fresh water has always been a popular target in warfare: bodies get dumped in wells, dams are busted. Conflict around fresh water is inevitable.

We can expect the conflict to get worse, because rivers are remarkably easy to monkey with. You wouldn't think so to look at one, because the force of all that moving water is palpable. Even an insignificant mountain stream, the kind that doesn't make it onto maps, has the power to topple a hiker attempting to ford it, buckling his knees, catching at his backpack, flinging him end for end in water that's only a metre deep. Nevertheless, people monkey with rivers all the time, big ones, damming them and diverting them. Walking a stream with a biologist who knows the local territory is a revelation, like going on one of those beneath-the-sidewalks tours of a great city. That attractive riffle where the family is camping? It's man-made, built ten years ago to take pressure off another eroding bank. The burbling stream you drove over on the way here? Its culvert under the bridge is too high and it's ruined for salmon. Man can take a lake the size of Ireland, fiddle with its inflow and outflow, and kill it. It happened to the Aral Sea in Kazakhstan, in the name of irrigation. Irrigation is also the reason the Nile often dries up before it reaches the sea. As does the Colorado. If the politicians and the engineers have their way, the same thing will happen to the São Francisco.

You can't get off the water cycle

In the part of the world where I live, fresh water is never far away. Often you simply follow it, a river winding beside the highway. Or you cross it, rumbling over bridges with little signposts telling you which river you're driving over. The names tell you something about the water: in coastal British Columbia, for example, Snowshoe Creek and Skim Milk Creek race heartily through the forest. Farther west, in foothills country on the other side of the Rockies, you're in a land of grain elevators, cattle ranches and oil pumps nodding lazily against the sky, and you cross trickles called Tongue and Risk.

I remember the point in my life when I looked out the window of the family car and realized that rivers flowed downhill, and I still enjoy the simplicity of the water cycle. With the exception of some wetland systems like the Pantanal or the great marshes like the Okavango in Botswana, it goes like this: snow melts, the water runs downhill and collects in rivers; rivers empty into the ocean; water evaporates from the ocean, floats around in the atmosphere and condenses eventually to fall on the mountains as snow. The beauty is the finiteness of it all — if you ignore the water that's trapped underground, it's just the same old molecules going around and around, year after year. Prince Charles could have skied over the water in the glass I'm holding now. We've got our water budget, and it's not about to change. It's just a matter of who's spending it.

The water that cycles from mountain to river to ocean and back is just part of a fixed planetary water supply that also includes groundwater, subterranean aquifers and the polar ice caps. Ninety-seven per cent of total global water is in the oceans. That leaves 3 per cent as fresh water, and the readily usable stuff in lakes and rivers is a mere 0.26 per cent of *that,* with the rest locked up in so-called permanent glaciers and ice caps and hard-to-reach underground aquifers. Water races downhill in rivers (which renew themselves every couple of weeks) or dawdles through lakes for decades, or collects underground in aquifers that may be prohibitively deep to get at. It may even sit in marshes and not go anywhere at all.

But it's not all geology and biology. Geography and its unholy handmaiden, politics, lean heavily on the water cycle, because the global freshwater supply is not shared equally. Huge chunks of it are concentrated in a few places: Lake Baikal in Russia, for example, has a quarter of *all* lake-held water. Canada has a tenth of the world's surface fresh water but less than 1 per cent of its population. (Fly across the country sometime anywhere north of the prairies. You won't see many cities, but you'll need sunglasses to beat the glare off all those lakes.) The Amazon Basin, which is shared by Colombia, Peru, Venezuela and Brazil, accounts for 20 per cent of *all* freshwater runoff.

Brazil itself has some peculiar water statistics. Of all the world's nations it has the most fresh water (about one-fifth of the total), but most of this comes from the Amazon Basin. However, Brazil is a lot bigger than the Amazon. Put another way, 93 per cent of Brazil's people have to make do with 30 per cent of the country's overall water supply. On average, the country actually provides more than the magic number of seventeen hundred cubic metres per person per day commonly used as the measure of an adequately watered population, but there are parts of Brazil — large parts — where it hardly rains at all. Here, people have always survived by living near rivers that originate in other ecosystems, by trapping water using small-scale, traditional methods like dams and cisterns, and by growing just enough to survive. In Part Three, we'll go to some of these places.

The key thing to remember is that the total amount of water is fixed, and that demand is increasing. What's happening in British Columbia, in the towns of Invermere and Tofino, is just a tiny example. Invermere, in the BC Rockies, is the last place you'd expect to have a water shortage. It's surrounded by frozen water half the year, but like so many similar towns it's been going through an explosion of development as retiring baby boomers flood in with their plans for six-thousand-square-foot homes with hundred-gallon water tanks, Jacuzzis, SUVs to wash in their driveways. Invermere isn't the only place this is happening; there are dozens of places like it in the mountains, hamlets that dozed for decades after the coal seams ran out, only to be jolted awake by an invasion of property buyers with a sudden need for a place on the slopes.

But the simple fact is that climate change has shrunk the glaciers, precipitation is coming more in the form of rain than as snow, and the snowcap itself is melting earlier in the year. Sinking new wells into the underground aquifer is one fix that's being promoted, not just in Invermere but all around the world, but there's a limit to the number of times you can do that, because aquifers need recharging. So, in the late summer of 2005, Invermere's town council put a temporary cap on all development, and the thought of such restrictions becoming permanent is hardly good news for developers.

Tofino is another working town that's turned to tourism, this time in one of the wettest spots of the rainforest coast of British Columbia. In Tofino, people simply forgot to plan for the water needs of all those summer visitors. In the late summer of 2006, bottled water trucked over the Sutton Pass was the only thing that kept the hotels open. The fact that all those palettes of Evian and Purely Canadian went jiggling past a very large lake just before reaching Tofino meant this wasn't a real crisis, just an ironic example of how water use can outpace planning.

Both examples remind us that the volume of fresh water is fixed. It cycles between rivers and lakes and clouds and oceans, but the total never changes. You can't actually take fresh water out of the water cycle; it's not like oil, which is extracted and used up. Whatever you do with it, it's going to find its way back to the mountains. If you use the water to irrigate your fields, for example, it'll still end up in the atmosphere, just not by the route it was expecting to take. Greater water demands, either from more people or because of new kinds of uses, mean the limits will eventually be pushed. By 2050, half the world's people will live in countries with chronic water shortages.

The biggest user of fresh water is irrigation in arid or semi-arid areas, but irrigation is wasteful: only a fraction of the water actually gets to the roots before it runs off or evaporates. How you add the water to the soil makes a big difference. Flooding or spraying the crops means that any salts in the water get concentrated in the upper levels of the soil, then percolate down into the groundwater. If there isn't enough rain to flush the salts out, the groundwater and the soil start to get salty, and eventually neither is usable.

Industry is the second biggest user, and tends to pollute the water before it's returned to the cycle. Domestic water use — for cooking and bathing — is actually quite small, around 8 per cent of the total, and nations vary hugely in what their people actually use. Canada is one of the worst spendthrifts; in summer, half of our water goes onto lawns. Japan is profligate too; Miss Tojo, I feel certain, leaves the tap thunderingly open while she polishes the stove or brushes her teeth. But domestic water may not be clean, and such a simple thing as dirty drinking water carries a frightful price in human lives: 1.7 million deaths a year, most of them children fatally dehydrated by diarrhea.

Until recently, water has been easy to get at. Unlike the oceans, which are hotly contested for what they contain but not for the water in them, rivers and lakes are there for the taking. All you have to do is stick a pipe in and pump away. But you can go too far. Rivers, big ones, can be sucked dry. And if there are other circumstances, like climate change, that reduce the amount of snowfall in one area and send it somewhere else, a region can suddenly find itself facing a water shortage (that's what happened in Invermere). Back in the old days, the easiest water was surface water and it was extracted using gravity, by building small dams and aqueducts and simply directing the water where you wanted it to go. Small dams are a great way to make use of limited water resources, although you need a lot of them, and governments are much more attracted to building large ones that entrap not ponds but lakes — oceans! — of water.

The water that's stored underground in aquifers is got at differently, by digging wells. Underground water is part of the cycle too, but it moves much more slowly than surface water. Aquifers also recharge slowly because water has to migrate through layers of soil and rock. The shallower, more accessible aquifers have already been tapped, so countries are digging deeper. Some of the aquifers now being penetrated are thousands of years old, like the Nubian sandstone being mined in Libya. Water is also being extracted from other huge reserves, like the Ogallala Aquifer in the United States.

Aquifers recharge faster in areas where there's lots of rainfall, and the water table (the depth at which you strike water) may be no more than a few metres down. But where rainfall is low, the water table can be hundreds of metres from the surface and, when too many drinking straws are sunk into any aquifer, the table falls, often to the point where all the water is gone or it's too expensive to dig

Brazilian reservoir fisherman: living off a man-made ecosystem.

any deeper. Lots of bad things can happen then, in addition to the region being useless for farming: rivers can shrink, or sea water can actually backflush into the aquifer, a real threat in coastal areas.

Around Mexico City, the water table has dropped twenty metres in the last fifty years, with parts of the city sinking after it. In India, pumping groundwater up to the fields has allowed huge increases in agricultural production, but much of the country is now paying the price in water shortages and contamination of coastal aquifers with seawater. In some places the water table is dropping six metres per year. It's the human species as parasite, sucking until the host is too weakened to feed it.

But real parasites are smart; they've evolved multiple life stages that burrow or swim or fly to the winds to emerge later on a completely different host. Humans may simply dam and drill and suck away until the globe runs dry. And then we'll plop off into the dust.

CHAPTER 5

TORMENTS
OF THE DAMMED

Fish meets concrete

In the shadow: fishermen launch a cast-net below the Três Marias dam.

"Brazilians think big," Uwe Schulz turned and shouted into my ear. He geared down and the little Fiat shot suddenly onto a narrow causeway that extended, arrow–straight and as far as I could see, over an expanse of water whose size and suddenness took me completely by surprise. Uwe jabbed a finger at the lake we were rattling over and shouted again, "And when these guys build dams, they *really* think big."

This was in 1996, before Três Marias, before the São Francisco spilled over into my life. Uwe was an expatriate German with steel-grey hair cut short and a meaty face out of Durer. It's almost impossible to separate the expats from the native Brazilians based on appearance alone, because so many Brazilians come from European stock. Uwe might easily have been native Brazilian because his Portuguese was flawless (so was his German and his English) and his wife was Brazilian. He had a faculty position at one of the Brazilian universities; I was making a precarious living as a conservationist who had somehow got invited to Brazil to help with migratory fish. He was driving me around the Grande River basin because his research program had an interest in preserving fish genetic diversity, and that was my specialty. Dams were apparently a problem. I wasn't clear on the details, but that's what this trip was for, to educate me enough to be useful.

Lack of sleep wasn't helping. I had arrived in Belo Horizonte the afternoon before, after twenty-three hours of travel from Vancouver. The next six hours were spent in the back of someone's car — I wasn't sure whose, although it had a logo on the door so it was either university or hydro company — swaying with the curves and dropping in and out of sleep. I vaguely remembered being shown into some kind of dorm with a half-dozen beds already occupied by snoring men, recklessly popping two of my magic sleeping pills, screwing foam plugs into my ears and sleeping right through breakfast (no papaya! no coffee!) and half the allotted time for the training course I was supposed to give.

I stumbled through it anyway, with Uwe translating "evolutionary bottleneck" and "minimum viable population" for my bewildered students, then signed formal training certificates for a string of baroque Brazilian names: Willibaldo Gallum, Eduardo Shimoda, Fabiana Cristina Silveira Alves de Melo. Then we all headed off to inspect the dam.

Now we were crossing the business end of the Volta Grande reservoir, one of dozens in Minas Gerais State. I'd never been to Minas before, only São Paulo. On the left side of the car, an inland sea of glassy blue water, serene and spreading; on the right, a thirty-metre drop to ugly brown whorls and turbulence. Under our wheels, the dam. The power-generating station was almost at the other side of the causeway, and as we rolled into the parking lot on the spillway, Uwe said, "And remember, this is only a little one!"

Look at a map of South America. Brazil and Venezuela are decorated with squiggly blue Rorschach blots intersecting the faint blue lines of rivers, as though the cartographer's hand was seized and shaken. On the map, a Brazilian river looks like a snake that swallowed several hats. The hats are hydroelectric reservoirs. Brazil, for example, has few natural lakes, unlike so much of North America, which seems in some places to be more water than land. From the air, the reservoirs are aberrations, their bizarre shapes giving the game away.

The first really big reservoir I flew over was the Guri Reservoir in Venezuela, a map-squiggle in the Caroni River: at an altitude of 4,500 metres the de Havilland took twenty minutes to grind over it. Approaching from downstream, I saw the dam and power plant first, a concrete stopcock outlined in froth clearly visible from almost four kilometres up. Then, behind the dam, the rapidly

Dam spillway, Furnas, Brazil.

widening expanse of the Guri reservoir itself, its boundaries chaotic and unnatural-looking, arbitrary as spilled ink. Narrow channels connected lakes two hundred, three hundred kilometres long. As the plane crawled overhead, the sun painted the water a dazzling silver. Up close, a dam impresses by sound and movement, and you can't help being thrilled by the sheer audacity of it. But from high above, different emotions take over, and you silently shake your head. An undammed tropical river doubles back on itself as though undecided, taking its time, an artery nourishing the land it meanders through. A dammed one is forever altered, hemorrhaging in one place, strangling in another.

The Grande River flows through the southwest part of the state of Minas Gerais, joining up finally with the Paraná. Volta Grande was the first dam I'd actually visited, and back then I had no idea how fortunate I was to be taken behind the scenes. From where we had arrived in our car that morning back in 1996, Volta Grande was very much sound and fury. We parked on a concrete deck that fronted the turbine buildings, high above the spillway. The deck throbbed underfoot and the noon sun was merciless. I'd had nothing to eat but a piece of watermelon salvaged from the missed breakfast, and only strong sugared *cafezinhos* to drink. (In Brazil, almost anything could be made into a sweet: *dulce de leite, dulce de cajú, dulce de laranja*. I wondered when the sugar would start oozing out of my skin, like sap.) Fifteen metres below the guardrail, the water crested and rolled in four huge, near-perfect circles — discharge whorls from the plant's four turbines beneath us. It almost looked inviting.

Authority announced itself in the form of the plant manager. He was courteous, uniformed and tired-looking, like someone resolved to maintain appearances after an all-night argument. No photographs, please; tours like this one were clearly by special arrangement. He led us across the deck and into the hangar-sized plant. Inside, a football field of concrete, with four embedded Mixmasters: the turbines. On this level we could see the tops of their shafts only, or rather the four identical green houses that enclosed them. Later I found out that each of them was capable of generating ninety-five thousand kilowatts, which made Volta Grande a "medium-sized dam." The throbbing underfoot was more visceral, and there was noise now too, and the smell of ozone. The concrete field — and it really did look like a field, since most of it was bare and painted in grids and paths — was walled with gauges and control panels.

There was a sense of furious energy leashed; only the dials moved. These were curiously old-fashioned-looking, as though lifted from a Second World War submarine movie. Not a digital gauge in sight. I learned later that Volta Grande was one of the older dams in Brazil (it was completed in 1974). The manager led us upfield, heading for the twenty-five yard line and the nearest of the green turbine-houses. A short flight of metal stairs opened abruptly onto a catwalk and we were suddenly confronted by the naked shaft of the turbine, burnished and howling in its yoke.

Finally I started to wake up. I remember feeling overwhelmed. My mind flooded, embarrassingly, with clichés of progress and industry and the harnessing of nature. I had no difficulty understanding the look of exultation on the politician's face when the spillway of a new dam is first opened. For this really was *it:* raw power, up close and in your face and in your nose and your ears and coursing up through the soles of your feet. The place fairly reeked of power, a machine-shop stink of burning grease and heated metal. Outside, the dam itself had seemed gigantic but comprehensible; water backs up here, *okay,* flows through some pipes or other under our feet, *fine,* turns some wheels and shoots out there, where those whirlpools are. *Got it.* But down here, trapped with the turbine, you felt like a mouse dropped into a crankcase.

We clutched the guardrail and the shaft roared at us. I wondered how many people had fallen into one of these hellish contraptions and whether it would mean a brown-out in Belo Horizonte or whether the lights would just dim momentarily, the way they do when you switch on the hair dryer at home. Since that visit to Volta Grande I've been inside other Brazilian dams — bigger ones, smaller ones, older ones like the earthen dam at Três Marias, and the wonder of the world mega-dam at Itaipú, so big its spillway looks like a sagging airport runway — but this one had the greatest impact on me. Because there was a sequel to this little demonstration of power, and it was waiting for me outside. In the river.

I was the last to leave. Back on the deck, cameras were out and clicking, but so far as I could see they were pointed at nothing, at a backwater beneath us near the foot of the dam.

"Over there, under the trees," Uwe said, lowering his telephoto and shepherding me toward a better vantage point. *"Curimatá."* I wasn't sure what *curimatá* was, although it was safe to assume it was some kind of river fish. Everybody here was a fish biologist, what else would they take pictures of? I squinted at the trees, expecting one or two fish, maybe a lucky glimpse of a dozen.

But there were thousands. They were most easily spotted near the shore, where the water was shallow, but when I knew what I was looking for and my eye began to add the subtle turning shape to the obvious flash of silver, I began to see a school of shadows. It drew my eye away from the bank, out below the spillway, until I realized the entire bay was filled with fish. They milled just beneath the surface, fat and ripe-looking. Now and then, one of them would turn on its side and arrow suddenly up, flashing silver. Whorls of them formed and dissolved. I scrambled down to the rubble at the water's edge and peered hard at them, eyeball to eyeball.

There's something about large numbers of animals concentrated in one place — herds of caribou, flocks of sparrows or penguins — that turns the assemblage into a superorganism with a will of its own. But the swimming of these massed *curimatá* seemed purposeless, without direction. They were big, some of them a metre long, but the most obvious thing was that they were going nowhere. The second most obvious thing was the tonnes of concrete nearby. Even my barely functioning brain, which I imagined shrunken to the size of a raisin and rattling around inside my overheated skull, could get that far. Hugo Godinho, the biologist whose cool, quiet house in Belo Horizonte later became my decompression chamber whenever I returned to Brazil, had spent his whole career studying these fish; he joined me now on the rocks.

For a while we busied ourselves photographing what looked like an aquatic holding pen. I snapped and pondered. It had rained last night, and the day before. It would probably rain again today. The river was rising, and the *curimatá* were on the move, heading upstream to spawn. Except that these ones had run nose-first into the dam.

"Hugo," I asked, "What's going to happen to them?"

That was it, my Big Question. And you'd think it would have been answerable. After all, this seemed pretty basic: dam in the middle of a river, fish frustrated on their way somewhere else, been happening for decades with migratory corridors for wildlife all over the globe and not just for fish, inevitable result of progress and the need for power. *Somebody* must know what's going on, right? The visitor from Canada with his Fraser River frame of reference was just asking another rather stupid question.

But Hugo didn't know. Neither did any of the others. There still isn't a satisfactory answer. What he told me then was the probables: some people thought the *curimatá* might turn back and find a side channel and spawn there, others thought they would fall back, try again next year. Back then, speculation and informed guesses were all anybody had to go on. The intricate route map of the dozens of Brazilian migratory fish species, now slowly getting filled in with

Waiting for a bite below a Brazilian dam (lights in background).

satellite tracking and painstaking collection of local information, was then little more than a few not-too-confident scratches on a big blank canvas. A simple question like the one I had just posed wasn't simple at all. While admirably suited to application of the scientific method, it was agonizingly complex, and the answer would affect not just the frustrated fish but everyone using the river — including the pourers of concrete. Everyone at least agreed on one thing: these fish would never get where they were programmed to go, to their spawning lagoons another hundred kilometres upstream.

We piled back into the Fiat waiting in the shade of the turbine building and headed off somewhere, I can't remember where, although by now it was midafternoon so I'm sure it involved food. It didn't matter, because I was going away with an image that finally gave purpose to nearly two decades of a haphazard existence as a biologist. Maybe "purpose" is the wrong word, because I wasn't suddenly fired up about saving the fish or dedicating my life to the design of better dams. Nothing like that; others are much better crusaders. It was probably the elemental quality of the image: blind, trusting, perfectly adapted nature confronted en masse by something a million years of evolution had never prepared them for. Irresistible force meets immovable object, except that this would hardly be a fair fight. Talk about rooting for the underdog.

I don't remember what happened on the rest of the trip, but I do remember that shoal of *curimatá*. I've never forgotten it. Poor brainless creatures, milling about aimlessly, not even permitted the space to coalesce into a superorganism like the great glittering balls of anchovies that crowd together for protection against marauding salmon or the sky-filling clouds of starlings I once watched from the roof of the FAO headquarters in Rome, painting the evening sky. Unlike the anchovies and the starlings, those *curimatá* were screwed.

STRANGLING THE SÃO FRANCISCO

The Grande, where I had my first "dams moment," has only four dams. A better place to see the effects of dams on fish is the São Francisco, where there are seven. When Sir Richard Burton launched his raft near Pirapora in 1867, the São Francisco flowed the way you'd expect for a large waterway that ran through several ecological zones, especially zones where rainfall was seasonal and highly variable. Its flow rate zoomed and faltered with the seasons, between eight hundred and eight thousand cubic metres per second. When it really got swollen it boiled along at fifteen thousand. The seasonal floods were devastating (one of my favourite Burton images is his description of catfish swimming through chapels high above the riverbanks). All that energy was irresistible to humans.

The water was needed for agriculture to develop the northeast, and so the dams went in. Proceeding downstream they were Três Marias (1952), Sobradinho (1980), Paulo Afonso I-IV (1955-1980), Itaparica (1988), Moxotó (1978) and Xingó (1994). I've been inside the dam at Três Marias, and it's safe to say things have changed a lot since it started operating. Brazilians are very good at building dams, and a modern one is marvellous in spite of itself.

Itaipú, for example, currently the world's biggest, although soon

Penstocks at Itaipú. The car is driving over the turbines.

to be displaced by Three Gorges in China, is like a cathedral inside, with soaring Gothic arches of concrete through which its three thousand workers cycle on one-speed bicycles. Itaipú has twenty-eight elevators, and the "floors" are marked in metres below water level. You can't visit there without gawking at the statistics that seem to have been written by someone just back from a European holiday: three hundred and eight Eiffel Towers' worth of iron and steel, fifteen Channel Tunnels' worth of concrete.

Even though it's actually bigger than Volta Grande, Três Marias dam seems like the Model T version, something you could more or less figure out by wandering around tapping on gauges and tugging on levers. It feels simple and manageable, like an old car. Três Marias is an earthen dam, where concrete is the icing, not the cake, and there's a mountain not far away with a hole in it where the rubble was scooped out to build it. A faded display in the lobby lists statistics: a million bags of concrete, twenty-one billion cubic metres of water in the reservoir. The inner wall is decorated with stains where water has seeped through the concrete liner, ever so slowly, leaching out the calcium carbonate to dry on the surface in a fanlike pattern. There are only a few levels, and the control room is mostly old-fashioned gauges and clunky-looking buttons — a wall of them, yes, but the kind of place where someone handy, the kind of person who can, say, change the oil in his car, could probably figure out how to run the thing.

Outside, on the catwalk that caps the dam, it's simplicity itself: if you want to open one of the five sluice gates you just open a little metal box and push a big green button. When I was there, the reservoir was low, just over half-capacity, and you could stand on the causeway looking now at the reservoir, now at the river, without detecting much difference in height. It didn't seem that opening a sluice gate would have much effect. But when the reservoir is full, the water shoots out as though from a cannon, and at least one person has been killed that way. Another good way to die at Três Marias would be to tumble down the service shaft that leads to the intake gates. Each one has a caged metal ladder that seems to descend into the bowels of the earth.

Mostly, though, the dam kills fish, not people, if not by preventing them from migrating, then by stunning them in the turbulence from the turbines or supersaturating the water with oxygen. The trees just downstream of the dam are a birdwatcher's dream, filled with white egrets waiting for an easy meal, and the water itself is cruised by cormorants called *biguá*. Sometimes, if you're a dam operator, you just can't win: when the turbines are shut down for cleaning, the gate takes an hour to close and the tube can fill up with fish, tonnes of them, which suffocate when you power up again. Fish kills like this mean a big fine and a lot of angry fishermen.

Dams mean the end of *piracema,* the annual journey of the migratory fishes. The most obvious way they do this is by blocking migratory pathways; even I, half asleep and bored in Volta Grande, couldn't fail to see that. But dams' effects are not limited to the mechanical. They also change the quality of the water by trapping sediments in the reservoir; they change the temperature of the river; most of all they allow engineers to control the flow of water, eliminating seasonal floods and the temporary marginal lagoons where many of the migratory species actually spawn, and where their babies spend the first critical months of their lives. It was those vanished marginal lagoons the *curimatá* at Volta Grande were looking for. All of these things — the lagoons, the floods, the sediments — make the river more than just a convenient water conduit from Minas Gerais to the Atlantic.

With each new dam on the São Francisco, the flow diminished and became more predictable. By the time the last one, Xingó, was completed in 1994, the river was chugging along at seventeen hundred metres per second, and it hardly varied at all because every time it rained the engineers just tweaked the dams up and down the river. Overall flow was down a third; the peaks and valleys were gone forever. There are still fish to be caught between Três Marias and Sobradinho but, beyond Xingó, it's mostly over. The river is too regulated, and the only hope seems to be artificial floods to provide some spawning habitat, a fine example of using engineering to re-create what engineering took away.

Reduction in flow also means that the river becomes uncoupled from the sea. One of the functions of rivers is to deliver nutrients to the sea, but the chain of dams in the São Francisco now snags and concentrates material before it can get to the lower reaches. Sediments drop out prematurely, in sandbars or in the enormous settling ponds that are the reservoirs; where the trees along its banks have been chopped down the river eats at its own boundaries.

That's the river now: stumbling if not on its last legs, seven times dammed, polluted and deforested, arriving weakly and predictably at the sea. Sometimes the experts come up with colourful language despite themselves — one report called the São Francisco "transparent and impoverished" — but nobody's listening to them now. Now it's the engineers and the politicians who stand on the crumbling banks and look down, their calculators out and clicking. Just a little more water . . .

What would Sir Richard Burton have to say about the São Francisco now? So far, we've heard from him only fleetingly, a footnote here, an observation there. But what this extraordinary man recorded during the short period he spent in Brazil deserves more than the occasional reference; behind the Mephistophelean beard

Três Marias dam, Brazil.

and the take-no-prisoners attitude he brought to his public life, Burton was a genius, a storyteller and a prophet. The future he foresaw for the São Francisco and the folly that has been made of it in our century are historical lessons that we ignore at our peril.

At the beginning of his voyage down the river, Burton wrote, "With a flush of joy I found myself upon the bosom of this glorious stream of the future . . . I had seen nothing that could be compared with it since my visit to the African Congo. In due time the banks will be leveed, the floods will be controlled, the bayous will be filled up and the great artery will deserve to be styled *a coelo gratissimus amnis.*" (The Latin, which he lifted from Virgil, means "beloved of the heavens are rivers.") If we heard nothing more from him, we might be tempted to dismiss him as little more than a Victoria booster, but that would be a big mistake. It's time we met the man properly.

CHAPTER 6

SIR RICHARD'S GHOST

The Kama Sutra and other interests

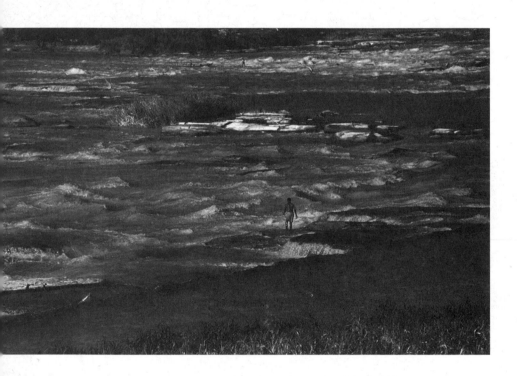

Hanging on by your toenails: fishermen work the rapids at Pirapora.

I FIRST RAN ACROSS SIR RICHARD BURTON at a time when I was hungry for heroes. In 1971, the BBC made a dramatic miniseries called *The Search for the Nile*. In one of the early episodes, long before "Dr. Livingstone, I presume," there was Burton slogging through the East African jungle. It was 1858, a time when European explorers seemed to be everywhere and discovery was a popular passion. Charles Darwin had made his world-altering voyage to the Galapagos Islands and Tierra del Fuego just twenty years earlier, and Burton himself would visit two more continents before he died. Now, in Africa, he was wrangling bitterly with his co-explorer, John Speke.

The producers of the TV show made no bones about the contrast between the two: Speke, the driven, tweedy Englishman more interested in shooting game than fitting in with the locals; Burton, the "dangerous" explorer and linguist who'd already been the first white man to penetrate Mecca (disguised as an Arab trader) and who filled his notebooks with every scrap of information he could gather on Africa's plants, geography, languages and ethnography. Speke was effete and constipated-looking; Burton was dark and gaunt, with an aquiline nose, high slanting forehead, deep-set eyes and jutting chin. His face was famously scarred by a Somali spear. What twenty-five-year-old in his right mind would *not* throw his lot in with such an antihero? When Speke took dastardly advantage of Burton's fever and hared off to track down the huge body of water he named Lake Victoria, I was rooting for Burton all the way.

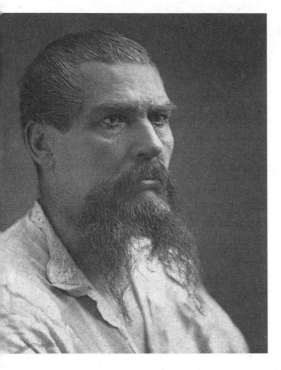

Sir Richard Burton, around the time of his sojourn in Brazil.

It got even better. After the return to England and a flaming controversy over the real source of the Nile, at the height of which Speke either shot himself or tripped over his own fowling piece (either way, he died), Burton was proven wrong. Lake Victoria really *was* the source of the Nile. My new hero had feet of clay, which made him even more appealing. Alan Moorhead, in his book *The Blue Nile,* called the explorer "one of those men in whom nature runs riot." Each of his more than twenty-five books is like a tropical botanic garden, an efflorescence of observations and ideas so dense and dazzling you constantly have to remind yourself that everything came from notes taken by a single person. Reading Burton, you lay the book down time and again and wonder: how on earth can he have seen all that, recorded all that, *done* all that? Burton made the most of every moment.

I started to read about this man. Not, at first, in his own words, the translations he was best known for, the *Book of the Thousand Nights and a Night* and the *Kama Sutra of Vatsyayana* — throbbing *lingams* and plow positions were too much of a distraction. In fact, most of his twenty-odd books about his travels were still difficult to obtain, so that huge body of work was unavailable. But there were a few biographies, and they painted a picture of a restless, overbearing iconoclast forever titillating and offending late-Victorian society with his tales of human skins (female, of course) brought back from Zanzibar, of impersonating a true believer in Mecca when discovery would have meant certain death, of storming out of a tent on the sandy Somali shore and doing midnight battle while his companions turned tail.

Burton never seemed to have made much money, living hand to mouth on a series of vaguely insulting government appointments and the revenue from the tales of his travels that he encrusted with every conceivable factoid. His gifts as a linguist were legendary: he would learn a language in weeks, and had more than thirty under his belt by the time he died in 1890. Those biographies made it clear that Burton was no one's apologist and that organized religion left him cold; and they were universal in condemning his wife, Isabel, as a hectoring Catholic who famously burned the only copies of *The Scented Garden of the Sheikh Nefzaoui* the day after he died. The image of portly weeping Isabel shredding phone-book-sized pornographic manuscripts by the backyard bonfire is a favourite in Burton lore.

So when I ran into Sir Richard's ghost on the São Francisco River, it was like bumping into a disreputable uncle you hadn't seen for years, someone the rest of the family made clucking noises about. What was someone of Burton's stature doing on an obscure river like the São Francisco? If he found himself in Brazil, surely he would have headed, as most travellers still do, for the Amazon? After all, making a beeline for the Amazon isn't a recent habit confined to socially conscious rock stars and development agencies (a foreign-aid representative once told me, "Stick a toe in the Amazon and we'll throw money at you"), but even in Burton's era the appeal of the place would have made any books that came out of a trip that much more marketable.

Part of the explanation comes from more recent looks at Burton's life and particularly at his finances: the title of the first biography I read about him, *The Devil Drives,* might better have been *The Bank Called and Your Last Cheque Bounced.* Burton and Isabel were chronically strapped, and his tetchiness and refusal to suffer fools made it hard for Victorian England to recognize his achievements. Burton was a polymath, expelled from Oxford, an unlikely combination of adventurer and scholar who is all the more appealing, to me at least, for having been at heart unable to choose one over the other. His contempt for bureaucracies made him an unending headache to the establishment.

The newer biographies spend more time on Isabel, the woman who had to live with this cantankerous genius. While she still comes off as the one who had to, in Burton's words, "pay, pack and follow," she also emerges as someone who, having scandalized her family by marrying him in the first place, proved herself physically as tough as he was, and just as capable of putting on pants and living on horseback while her husband collected stories and plants and bits of rock. When Isabel injured her leg and was unable to continue on the water portion of Burton's São Francisco trip, she rode back to Santos unarmed. It was a journey of more than six hundred kilometres, with only one manservant to accompany her: Chico, a dwarf the Burtons called their "right hand man in Brazil."

Burton spent his twenties in India, in the army of the East India Company, where his proficiency in the local languages made him a natural government agent. Later, he had himself circumcised and travelled incognito throughout Arabia. At thirty-three, he was serving as a British agent in the Horn of Africa, often disguised as an Arab merchant, and the book he wrote about his adventures, *First Footsteps in East Africa,* is so readable and comprehensive it's still the standard text on the Somali society of the time. It was on that trip that he acquired the famous scar, when he was attacked by a warrior on the coast, the spear entering one cheek and stapling his jaws together.

Burton was protean, penetrating Tanzania without maps, following the few tracks of animals and slave traders all the way to Lake Victoria. Others have hacked their way through similar horrors, and modern adventurers still do, but they're often accompanied by a film crew. None will ever again come close to the prodigies of examination and interviewing and note-taking that were Burton's daily routine.

But while he was practical and meticulous he still managed to be poetic. When, toward the end of his journey down the São Francisco, he finally reached the spectacular falls at Paulo Afonso, he wrote: "Presently we heard a deep hollow sound, soft withal, like the rumbling of a distant storm; but it seemed to come from below the earth; after another mile the ground appeared to tremble." This was the "Niagara of Brazil," a seventy-five-metre gorge that choked the river to fifteen. "It is filled with what seems not water, but the froth of milk," Burton went on. Experimenting as always, he hurled stones into the chasm; each was stopped dead after six metres. Enchantment, said the guides; wind, said the explorer. He called it "power tremendous, inexorable, irresistible." Finally he says, "I left the place that the confusion and emotion might pass away."

This is quintessential Burton. So too is what he said about an eroded bank of the São Francisco that obstructed navigation: "the material is lamellar shale, porous, and full of holes; it might easily be removed by a small steam-hammer." Another person might speculate about removal too; who but Burton would give equal space to the geologic makeup of the obstruction? Above all, he could *write*. Here he is after camping out on a sandbar on the São Francisco: "the night was of a stillness so deep, an unprotected candle would have burned out."

Why Burton was in Brazil

In the mid-1860s, after Burton had finished a stint as Her Majesty's consul in Fernando Po (now Equatorial Guinea in West Africa), the government continued what would become a series of attempts to hide the Burtons somewhere harmless: they made him consul to Brazil. Politically there wasn't much Burton could do to jeopardize British interests in the country because, practically speaking, there weren't any. Britain had her colonies in Asia, Africa and the Caribbean, leaving South America to Spain and Portugal. Burton would be installed in swampy Santos near the Atlantic coast just east of São Paulo and would keep the flag flying. He was forty-six years old.

Burton stayed in Brazil from 1865 until 1868, and whatever "official" work he did as representative of the British Empire survives in reports and letters that are little different from what a lesser consul might have produced. He had the job down cold within a few months. He established a household in the healthier heights of São Paulo; he wrote and translated feverishly (Camoens' *Lusiads,* a collection of Hindu fairy tales, bits of the *Arabian Nights)*; and he developed a passion for the mining country of Minas Gerais. This led to two things.

The first was typical Burton: his interest in the gold diggings in Minas led to a purchase of shares, which led to an official complaint by the British ambassador to Brazil — a complaint that was squashed by the foreign secretary, Lord Stanley (he of "Dr. Livingston, I presume"). It helped to have friends in high places. The second development was a trip on horseback throughout the mining country of Minas Gerais, to be followed by a trip by raft and canoe all the way down the São Francisco River. Isabel went with him for the first leg, enduring heat on the plains and cold at high altitude, putting up with three months of dust, bugs, highwaymen and disastrously dangerous mineshafts. In her memoir, she wrote at length about descending, in a kind of iron kettle on a rusty chain, one kilometre into the Morro Velho gold mine, the deepest in Brazil:

> It was a stupendous scene of its kind. Caverns of quartz pyrites and gold, whose vaulted roofs, walls, and floors swarmed with blacks with lighted candles on their heads, looked excessively infernal. Each man had drill and hammer, and was singing a wild song and beating in time with his hammer. Some are suspended to the vaulted roof by chains, and in frightful-looking positions; others are on the perpendicular walls. . . . After seeing the whole of this splendid palace of darkness in the bowels of the earth, we sat on a slab of stone and had some wine.

I visited the same country by car from Belo Horizonte, and the hills are laid open almost to the road. You see great winding terraces cut into the sides of mountains, and the roadside stands sell pots made of iron and stone and a clay so dense your stew takes an hour to come to the boil and almost as long to settle down once you've turned the stove off. Around Ouro Preto the history is palpable; you can visit the Museo da Inconfidência that commemorates the independence movement or "Minas Conspiracy" of 1789, and see where the head of Tiradentes, the leader, was impaled on a pole after his quartered and salted body was brought from back Rio. I saw tables more than seven metres long and made from only two planks of *jacaranda*.

Fortunes were made here and lives squandered. In one restaurant in the old city you can finger the wooden yokes that went around slaves' necks before you settle down to a hearty meal. Now, Ouro Preto is mostly about having a good time, a place to eat excellent steak while listening to a barrel-shaped sixty-year-old woman belt out a seamless forty-minute set of Gershwin and Jobim, or order a Kamikaze cocktail in a jazz bar operated by a woman with a degree in botany and an inventory of a hundred different kinds of *cachaça*. (The Kamikaze is *cachaça,* lime and Cointreau, and it's deadly.)

Isabel would have run the river with her husband if she hadn't fallen and put herself on crutches. Back she went with Chico to São Paulo, and Burton did the São Francisco alone. It was this trip that the BBC's present-day blogger attempted to repeat, and it was Burton's book about it, *Explorations of the Highlands of the Brazil,* that made me rethink my first impressions of the place.

Meeting Burton's ghost on the São Francisco

I returned to the São Francisco a year or so after my first visit. The last-minute introduction Barbara Johnsen had given me had been like a preface; now I wanted to read the whole story. This time I drove myself from Barbara's town of Três Marias north to Pirapora, a bigger and more graciously situated place that sits across the river from its sister city, Buritizeiro. The potholes in this stretch are legendary, and a sudden opening of the heavens just before my destination turned them into little red lakes. Avoiding any of them was like playing a video game, except that the other player was a charcoal truck coming straight at me. Lanes became irrelevant, and there was always a truck in front of me too, blocking any attempt to pass unless I really trusted the rear turn signal code: right to pass, left to stay back. I saw two major accidents, one of them a car that had skidded into the rear end of a flatbed truck, the cargo of metal girders punching neatly through the windshield like a pin through an insect. Blown tires curled up on the shoulder like the skins of some enormous reptile.

As I neared Pirapora, the road wound back down into the valley and the São Francisco reappeared below, undulating and chocolate brown. Where the *cerrado* still existed, termite nests sat in the crotches of gnarled and scaly trees, as though flung there to dry in the sun. The *cerrado* here has always been cowboy country, and in the truck stops you can buy a leather ten-gallon hat for $30 or a *maté* vessel made from a cow's horn. The closer I got to Pirapora, the clearer was the process of the São Francisco's "development" — vast swaths of native *cerrado* had been cleared and replaced with tracts of eucalyptus and soybeans and brand-new vineyards financed by Japanese yen.

Soybean goes into the ground in October and is harvested four months later, and when you drive through the valley of the São Francisco it's not hard to believe that Brazil is the world's biggest producer. The country has invested heavily in creating varieties of soybean that grow fast in the tropics and get their nitrogen not from fertilizer but from symbiotic micro-rhizomes inoculated straight onto the seeds. There are places where it's like being on the Canadian prairies: all you see is green fields, blue sky, white clouds.

The *cerrado* replaced by these wonder crops is a world hotspot of biodiversity, with the richest flora of all the world's savanna ecosystems. Everyone knows the Amazon rain forest is being shorn — we have any number of celebrity conservationists to tell us that — but the rate of deforestation and the total area cleared are actually higher in the Brazilian *cerrado*. You'd think the rock stars would be drawn to the wide open spaces, the high plains, the cactus — in the *cerrado*, almost everyone seems to play the guitar — but they seem not to be.

I must have let my annoyance at Bono and Sting distract me, because I hit a speed bump going into Pirapora that nearly tore the bottom out of the rented car. I was in the thick of town before my heartbeat returned to normal. I found a room in the Hotel Cariris, a cavernous place with a diseased-looking pool on the roof and named after a long-vanished Indian tribe. But the room was fine, with a tiny tiled balcony that looked over the main street. And beyond, between the tiled roof of a beauty salon and a fibreglass water tank, I could see a sliver of river.

It seemed sometimes as though that was all that remained of Burton's Brazil, just a sliver. In Burton's time, the rapids at Pirapora were much more formidable than they are now. No sooner did he arrive than he began to figure out how to change that. He devotes pages of his book to removing the rapids and, while he's at it, building dams for flood control and making the rest of the river navigable. The estimates of previous German and French engineers he dismisses as inflated (and mercifully banishes to a footnote), but he goes on to provide his own, right down to the number of sledge hammers and picks. Burton is careful to preface his estimates with the disclaimer that he's not a prospective contractor to the Brazilian government, but you get the feeling he'd have liked nothing better than to jump in himself and start blasting — if only there weren't so many other things to do. Species and rocks to describe, people to meet, rapids to run.

The rapids of Pirapora are still there, between the towns of Pirapora and Buritizeiro, and even though much diminished now (someone did eventually dynamite the bigger rocks), they're still a fought-over fishing territory. Três Marias, of course, didn't even exist when Burton passed through, and it was impossible, when bucketing along that spectacularly potholed highway to Pirapora, not to realize that, even in the blinkered view you get from a moving car, the countryside had been catastrophically altered in the last century and a half. Burton's Brazil had vanished, and so had his vision for the São Francisco. The railroad he dreamed of, for example, was here as a relic in the form of the old bridge between the two towns. People still cross it on foot and by car, straddling terrifying gaps in the old railbed, but the trucks won the transportation battle long ago. In the eddies below the bridge, fishermen beach their canoes for temporary repairs.

Burton's journey on the São Francisco really began here in Pirapora; the month spent rafting the Rio das Velhas from near Belo Horizonte to Pirapora was simply the prelude to the main act. Regulation of the river means the famous rapids are no longer navigable by vessels of any size, but you can hire a boat below the worst of the rocks and chug downstream as far as the next big dam at Sobradinho, almost three hundred kilometres away. I found *Senhor* Denio, a fisherman turned entrepreneur and tour guide, and he took me partway there on the *Taina,* a squat thirty-footer outfitted with a comfortable shaded viewing deck, minimal cabins for six and a real bathroom with a hand-held length of red rubber hose for a bidet. Hanging over the prow was one of those carved, snarling *carrancas* that Germaine Greer, with pithy disparagement, called the "worst sort of debased folk art." As *carrancas* go, I thought this one got the job done, staring down any evil spirits coming the *Taina*'s way.

We boarded in front of the long promenade that runs along the river, a place for strolling and cycling or just standing still and letting the others move for you. I crossed a plank laid over the mud and when I looked back up at the town a man cycled past playing a guitar, closely followed by another cyclist reading a newspaper. It beat dodging potholes and charcoal trucks. Denio pulled in the plank and we got underway.

The *Taina* chugged downstream past experimental irrigation projects concealed behind a buffer of *gameleira* trees. A few kilometres from Pirapora we passed an industrial area where three refineries produced silicon from Minas quartz, all fuelled by eucalyptus. A stream of trucks brings fuel and ore in and carries product out, one reason

Silicon being refined from quartz, near Pirapora.

the roads around Pirapora were so execrable.

Other factories in the same area make textiles and jeans; all there because of the river, all beneath a cloud of smoke. Where there weren't factories there were cows, and farms growing manioc, corn, mangos, bananas, even sugar cane. All sucked at the river. What remained of the marginal lagoons was back there too, waiting for water that came less and less often as the river was ever more strictly regulated. The navigation signs were like nothing I was familiar with back home, on the ocean; one I kept seeing had a big X on a white background.

"Arreia," Denio said, miming a person walking into a glass door. It wasn't a word I knew, but I figured it out soon enough: the X marked a hidden sandbar that could stop a ship cold. Another sign read, "1976," the numbers faded almost to white. It wasn't a date; that was how many kilometres were left before we hit the Atlantic.

The Volvo diesel ticked over down below, letting the current do most of the work. The water of the São Francisco was the colour of milk chocolate. Denio and I watched it all drift by from plastic deck chairs while his assistant dodged the sandbars. He talked non-stop, machine-gunning facts and opinions at me until I had reached the point where Portuguese was noise, not speech. Denio was affable, helpful to a fault, an undoubted font of knowledge. But he was driving me crazy with his lecturing. I felt like a character in a Western movie, implacably pursued by a rider a hundred paces behind. I turned and lobbed a stick of dynamite at him.

"And what about Burton?" I said.

Denio's conversational horse burst through the smoke and came remorselessly on. "Richard Burton, of course." *Hichard Bear-tone,* it came out. "Of course, I know him." Not only knew him, but had actually read him: Burton's *Explorations of the Highlands of the Brazil,* it turned out, was easier to get hold of here than in the original English back home. We slid past two very old, very portly women in a fishing canoe. I had no idea how they could have got into it, let alone fish from it. If they hooked into a big *dourado,* one of them would have to jump over the side. Denio raised a hand to them without slackening his conversational assault and I threw in the towel. "Uh huh," I nodded, Yep, *claro.* The words washed over me, closed over my head. I was no match for Denio. Furthermore, he had read Burton's book and I hadn't.

Later, when I finally got hold of Burton's *Explorations of the Highlands of the Brazil* and waded into the São Francisco with him, I found out what Burton thought of the Denios of his time. He has high praise for boatmen, especially the aged pilot who took them to the final obstacle, the Paulo Afonso rapids, a man with a "queer dry humour, who delights in chaffing the people on the banks, twanging the guitar, and taking snuff." Like almost all his countrymen, the pilot had an amiable defect, a constitutional inability to say "No." Another crewman has an immense repertoire of songs, "drinks like a whirlpool and eats like an ogre, and is abominably good-tempered." These were rascals, ruffians, men in some ways on a level with Burton. His observations were the kind that come from total immersion, both cultural and linguistic. They made my hurried efforts at connecting with Denio seem laughable.

A LESSON FROM A FISHERMAN

The next day I went to see some fishermen and encountered Sir Richard's ghost once again. Fishermen in Brazil are organized in *colonias,* associations that look out for their interests, with a president, a bit of an infrastructure, varying amounts of clout depending on local issues and who's at the helm. Pedro Melo headed the Pirapora colony and he had his hands full. When I first met him he was arguing passionately with some fishermen from the colony in Buritizeiro, across the river from Pirapora. Buritizeiro gets its name from the *buriti* palm. Burton describes the dependence of local Indians on the tree: they used its fronds for baskets and fences, the fibres for hammocks, the juice for wine and the pulp to chew as a sweet. Pedro's problem this particular morning was about who fished when and where in the rapids that separated the two communities. It was a classic turf war of the kind that gets any fisherman boiling.

Mel means "honey," but Sr. Melo was being anything but sweet as he made his points by the river's edge. A bleached-looking pony wandered along the shore and, farther out, two young fishermen crept from one submerged rock to the next, nylon cast-nets slung over their glistening shoulders, inching from shore and safety to where the big, salmon-like *dourado* were running. The men were deeply black, like Pedro, probably descendants of the slaves who came filtering in from the coastal cane plantations in Burton's time, into the interior *cerrado* of Minas and Bahia. One false step and they would be under water in a tangle of limbs and shot-weighted nylon.

They were catching strange fish these days, things like *tucunaré* or peacock bass, a brightly coloured fighter introduced from the Amazon. *Tucunaré* were popular with the sport fishermen, but what effect would they have on the local species? Nobody seemed to care; maybe Brazilians just fundamentally embrace immigrants. It's too bad, because there are places in Brazil where a hand-net dipped into a river by the side of the road will yield more variety than the hand-bag department of the Sogo Department Store in Tokyo. Introduced species almost inevitably simplify ecosystems as rich as this.

When the argument finally concluded, the fishermen were just specks on an invisible rock midway between the warring towns. All I could see was the occasional glint of wet nylon, a tiny silver disc opening for an instant against the sun as the net was launched into the boiling water. The old women in the tiny canoe, these guys clinging by their toenails to a slippery rock half a kilometre out — there didn't seem to be much room for the catch in either case. Pedro led me back to my rented car and we headed the few kilometres along the river to Barra de Guaicui. (Pedro normally travelled by bicycle, a lanky figure I kept seeing weaving unhurriedly down one Pirapora avenue or other.)

Guaicui is a hamlet, near where the Rio das Velhas joins the São Francisco after travelling all the way up from Belo Horizonte. You might say the Velhas takes the direct route to Pirapora, not the roundabout highway drivers have to follow, and it arrives so contaminated with Belo Horizonte's sewage and industrial effluent that fishing isn't even permitted — not that that stops anyone. When Pedro and I turned down the road to Guaicui the only sign of life was a barge tethered beneath the tree canopy on one bank, sucking up sand through a noisy gas-powered vacuum hose the thickness of a man's waist. Also illegal, but in Brazil concrete is on tap like beer,

Two commercial fishermen near Guaicui.

and the sand has to come from somewhere.

Guaicui is the last place Burton stopped before entering the São Francisco, and his description is the usual blizzard of information, including the price of a *curimatá* ("of course, cheap"), and this description of the diet: "fish and manioc, manioc and fish." We passed a ruined church near the riverbank, where a thirty-metre *gameleira* tree straddles the remains of one wall. The root structure flows down over both sides of the stone like alien tentacles, thick and nearly black, and the space between roots is covered with graffiti. The church is the Bom Jesus de Matosinhos and it was abandoned, half-finished, even when Burton visited, "pilasters and pulpits of cut stone destined to remain in embryo." About the locals, Burton was even-handed: "when the men are sober they show nothing of the ruffianism of the European uneducated." As always, Burton has time to explore, and to meet hundreds of people, from peasant to prelate. And as usual, he has advice, this time for the local government: widen the streets for the days when "tramways will become universal."

The tramway never did come to Guaicui, and the fishermen haven't left, despite the mess others have made of their river. Amadeu, the man Pedro wanted me to meet, lived in a clay-brick house down a scarred and wrenching dirt road. The place was tiny, a single room with naked windows covered by floral cotton curtains. A dirt floor led past an enamelled stove under a plastic oilskin, and a small black-and-white television perched on a dresser beneath a carved Christ on the cross. An ornate wooden bed took up half the space, neatly draped with a cream-coloured lace spread. The family was in the back.

I was never introduced to Amadeu's wife. She was black, pregnant, unhappy-looking, in a stained pink top and a short white dress rucked up over her thighs. Her canine teeth angled forward and the ones between them pointed back, as though a fence were collapsing in her mouth. She stood silently beside her cooking area, an outdoor mud oven with a metal grate and a stack of scraggly branches piled beside it. Next to the stove a one-centimetre plastic pipe discharged a wavering stream of water into a shallow tin bowl overflowing into the dirt. The water vanished as soon as it hit the ground, and I kept eyeing it. After standing around on the banks of the river watching Pedro Melo argue for an hour, I was parched.

"Ruim!" Amadeu was small, wiry, shirtless. He was brown, not black, his chest absurdly furry. *Ruim* meant bad. Amadeu laid aside the long, single-bladed oar he had been carving and motioned to the water spout. His wife flapped into the house, reappeared with a glass, filled it, handed it to me. Barra de Guaicui was fortunate; it had a source of piped water, purified after a fashion. Amadeu grinned while I drank it down. His teeth were long and yellowed, with a big gap in front. He squatted, motioned to an old stool, began to talk.

Who was I, why had I come, and the fish — what was happening to all the fish? The gap between his teeth and the chopped-off dialect of the region made it hard to communicate, but we managed, paddling hard against the linguistic rapids to establish some essentials: his life, the state of the fishery, his daughter. Where was his daughter? Amadeu vanished into the house and returned with a three-ring album gloved in sticky clear plastic. There were only a half-dozen pages. One had a curling snapshot of a little girl, perhaps three years old, the colours faded and blurred from heat and humidity. He handed me the album, pointed to the picture, put his palms together beside one cheek then pointed up, all the way to heaven.

I thought about my own kids. I drank some more water. Pedro returned from the miserable garden patch with a manioc plant and handed it to Amadeu's wife. We all watched her silently hack the long tuberous root into oozing sections, drop them in a battered aluminum pot, snap an armload of gnarled brushwood into lengths for the fire. The process was agonizingly slow, and there was a lot of manioc in that pot. She used the bark first, for kindling, working through several precious lengths of wood before the water started to boil, deforesting one more little patch of native *cerrado* in my honour.

I wasn't even slightly hungry; all I could think about was the absurdity of my presence here, eating these people's food, making furtive mental notes about their house and their mean little collection of belongings, representing God knew what conception of another impossible world before I got back in my car and drove off in a cloud of dust. I felt like a development voyeur. Somewhere just beyond the edge of their manioc patch the Velhas River carried its cloud of pollutants to the São Francisco. Burton must have passed within a hundred metres of where I was sitting now, waiting for five kilograms of starch to come to a boil. I asked Amadeu, "Have you heard of Richard Burton?"

And he had. The English explorer, yes, of course. Amadeu nodded seriously, as though knowing about Burton were unremarkable, a given. A subsistence fisherman, illiterate, in a hamlet on a river that's essentially unknown to westerners, considered it a matter of course to be aware of an English traveller from the nineteenth century whose career is little more than a curiosity in his own country. Ask anyone in North America who Richard Burton is and if they're old enough to remember the sixties they'll say, "The actor? The one who gave Liz Taylor that honking big diamond?"

I ate the manioc, as much as I could get down; the steaming discs of potatolike tuber sat in a bowl and we fingered them out, dipped them in salt, chewed. I like manioc, especially fried, but today it clogged my mouth like glue and I had to take glass after glass of

Amadeu's "bad water" to wash it into my stomach. We left soon after, and when I returned to the hotel in Pirapora I took a beer out onto the tiny porch and slumped down against the outside wall. It was warm against my back, and the shiny brown tiles on the porch sent more heat up through the soles of my feet. Below, people went in and out of the pharmacy across the street but the rest of the shops were closed for the day.

I decided I liked Pirapora, with its view of the rapids and its air of experience. I liked a place where people rode bicycles while reading newspapers and still managed to avoid hitting one another; where a beauty in long red hair and silver hoop earrings could perch on her own bicycle seat and watch a game of soccer on a sandbar that Richard Burton might have passed; where a tiny shop sold old sewing machines, broken keyboards, the Beatles' *Abbey Road* and *Hits of Carnaval, 1964*. Pirapora was the kind of place where pace depended on your outlook: tiny ants raced busily over the sticky plastic tables while a dappled grey horse stood motionless for hours between the traces of its cart, waiting for orders. The beer was cheaper than water and the food was plentiful, as long as you didn't mind beans, eggs and deep-fried chunks of *surubim* catfish from the river. The feel of the place was entirely different from Três Marias, where the population is polyglot, the descendants of the people who came from all over Brazil to build the dam and the refinery. Pirapora, on the other hand, was more "northeast," less cosmopolitan, and the closeness to Bahia was everywhere visible in the number of black faces. Pirapora had history.

An old guy went slowly by, in grey slacks with a knife on his belt, and the last of the sun caught the rim of his cowboy hat. His sandals slapped on the pavement. If I craned my head against the railing I could just catch the São Francisco coursing past the town. Denio's boat was tethered out there; Amadeu the fisherman was only an hour behind me; both of them knew of the foreigner who had stopped here for only a day or so, a hundred and fifty years ago. That had to mean something.

What it meant to me then, with a cold can of beer fitted over one kneecap and my hair still wet from the shower that had been so easy for me to step into, was that the locals remembered Burton for

Tending bar in Pirapora, Brazil.

one good, simple reason: because he cared enough about their river to visit it. The São Francisco, Velho Chico, that muddy thing chugging along behind the rows of buildings radiating heat into the evening sky, with its armada of plastic bottles from Belo Horizonte and heavy metals from the refinery in Três Marias and its sewage from everywhere in between. It wasn't the Amazon and it certainly wasn't the Fraser I was used to, but the river seemed to matter deeply to everyone who lived near it. If I was going to keep coming here I would have to start caring about it too.

The problem was, who would I return as? I'd come first as a scientist with tools to apply to problems of the local fisheries. But already I could tell that, for people like Amadeu, science wasn't enough — might not even be relevant at all. What good was science in the face of a hundred years of pollution drifting down on the poor guy's fishing grounds?

I thought back to my visit to the dam at Volta Grande years before, when nobody could answer a simple question like what would happen to the thousands of *curimatá* butting heads with a concrete wall across their river. *Curimatá* were interesting to science and to Burton: "The naturalist who shall attempt the ichthyology of the São Francisco will have before him a task of years," he said, offering up the example of the *mandim* that "croaked like a frog and grunted like a pig under our bows." *Curimatá* was only one of a half-dozen local names for the same creature, and Burton lists all nine of them in a footnote, along with the notation "caught with hook and line, baited with pieces of banana."

Their fate at Volta Grande was a delicious scientific question, the kind biologists love, the kind a green graduate student would give his eye teeth for. It should be answerable, that's what the "scientific method" is for: dream up a hypothesis, then test it by collecting data that either confirm the hypothesis or shoot it down. In the case of the frustrated *curimatá,* the Brazilians already had some hypotheses to test. But what good was finding the answer if nobody listened? What good was *science?*

PART II

Science, Sex, Sushi

CHAPTER 7

"BACK OFF, MAN: I'M A SCIENTIST"

Somebody else's life

The science tourist.

THE BOTTLE ROOM WAS MY UNDOING. If I hadn't spent so much time in the bottle room I would never have been robbed at gunpoint on a deserted Brazilian beach by a person in a red bathing suit. Future generations are safer now because the bottle room is long gone, but it ruined my life.

When I was a kid, the city of Victoria had a fine old museum with a fine old name: the British Columbia Provincial Museum. That was what it was, a lot of things from British Columbia: rocks, cultural artifacts collected or cajoled or stolen from the Indians, insects and sea creatures in jars and, in one great hall, the big animals themselves. Wolves, bears black and grizzly, cougars, moose, all of them stuffed and mounted on ornate wooden pedestals and looking at me with big brown glass eyes. Except the wolf. He was grey, and twice my eight-year-old size, and his eyes were a cold cloudy blue. I loved the Provincial Museum. It was somewhere I could go by myself, it was never crowded, I could wander the halls in the great stone building without running into a soul. It seemed to be my own private place. My favourites were the bird room and the bottle room.

The bird room seemed empty at first, because each wall was nothing but shallow drawers, each with a number and a little brass handle, about the size of a couple of phone books laid side by side, row upon row, as high as my head. When you pulled on the handle the drawer slid smoothly out and inside it lay the stuffed bodies of birds, maybe a half-dozen, each with its wings held neatly to its sides and its feet tucked together, as though plummeting toward some hapless prey. On one claw there would be a small paper tag with names and numbers and probably other information, like where they were collected and by whom, but I was interested only in the birds themselves. I saw so many of them, the ones that were brilliant in life still brilliant, the barrel-chested divers still puffed up as though ready to plunge beneath the surface at a moment's notice, the eyes, like the eyes of the stuffed animals, perpetually brown and sparkling because it was never dusty inside the drawers.

The smallest drawers contained the tiniest species, like hummingbirds, a dozen iridescent specimens laid out like jewels. The biggest drawers held only a couple of birds the size of cats or even dogs: herons, loons, mergansers, eagles. If someone had walked in and handed me one of those birds, even an insignificant little one like, say, the black and white bufflehead ducks that bobbed around Victoria's waterfront during the winter, my life would have been complete.

The bottle room seemed less popular than the bird room; I remember always being alone there. This may have been because of its smell. Apart from the jars themselves and the pickled, pallid forms they contained, most of the room was filled with formaldehyde or alcohol. If I had been of a mathematic turn of mind, I could have calculated the amount of noxious fluids on all those shelves; probably it was several swimming pools' full. But I had no aptitude for math (another career warning sign ignored); I was captivated only by the creatures themselves. If the bird room and the stuffed animals on their pedestals were about what lived on land and in the air, the bottle room took you beneath the surface of the ocean, where the wonders were. A decade later, Captain Cousteau would invite television viewers down where these creatures dwelled and create the mini-boom in marine biologists of which I was briefly a product, but all I had then were the bottles and my imagination, and that was enough.

My mother, who possessed a kind of sweet indefatigability for divining and promoting the interests of her children, soon cottoned on to my fascination with the bottle room. Arrangements made by adults proceed wholly unobserved by small children, so I was surprised when I found myself in the bottle room one day, not alone, but in the company of my mother and a long, lugubrious gentleman with an untidy rooster tail of white hair. This was Clifford Carl, Dr. Carl, the director of the British Columbia Provincial Museum. Why a doctor should be running a museum I didn't know. Mostly I remember him leaning over me, and wondering when he would pull out his stethoscope. I do recall his saying, at the low point of one of those bends, "Well, I suppose you'll grow up to be a biologist."

And I, not wanting to disappoint, said yes.

Dr. Carl went on to write books on the flora and fauna of British Columbia that are used to this day, but his eclectic museum was turned into a cultural entertainment emporium filled with interactive displays and fibreglass recreations of critical ecosystems that left me cold. The stuff that used to do it for me, the birds and the bottles, are still there, archived in storerooms open only to researchers, the plumage of the entombed birds slowly fading but the pickled sea creatures unchanged from the day they were tagged and immersed. The public will never see the bottles again.

It took me half a career to realize what went wrong during my interview with the kindly Dr. Carl. We both made the same mistake: that fascination with the contents of all those bottles and drawers meant suitability for the process that put them there. For example, the fact that there are a hundred varieties of nudibranch off the shores of British Columbia is unquestionably a marvel. In an earlier age, when there were still amateur biologists, one could quite reasonably have made the study of nudibranchs a diverting and lifelong hobby, the stuff of basement laboratories and sudden, middle-of-the-night floods. But we don't have amateur biologists any more, or amateur anythings, so the only thing for me to do was to become a professional one: a scientist. Science was the way things got into jars. An interest in nature, a career in science, it all made perfect sense. I am sure all three of us went home that day feeling that at least one tiny human problem had been solved.

Going into science based on an interest in the wonders of nature is like deciding to become a musician because one likes music. Musicians don't just get up and play; being a musician involves a lot of gruntwork, the dogged repetition of a troublesome passage until it sits securely in the fingers. It's the equivalent of the scientist's persistent collection of data that will, in the end, make or break his argument. Most music lovers are a lot better off just listening to music; most nature lovers should stay away from science and just enjoy nature.

I didn't know any of this when, twenty-odd years after Dr. Carl's fateful pronouncement, I started out on a career, fresh Ph.D. in hand. All I knew was that I didn't want to spend my life endlessly refining one trivial theory of how one infinitesimal part of nature worked, so I took my scientific training and my door-opening degree and became an applied scientist. And not just any kind of applied scientist, but the worst kind: I became a science tourist.

A Ph.D. from "outside" goes a long way in developing countries. Because science is a big part of international development aid, the science tourist takes his or her training and reputation and applies them to "helping out" in another country. Some scientists make forays from university bases back home, some stay on in the developing country as consultants or administrators for the development agencies themselves. The very best are motivated by a real passion for helping and choose their destinations on the basis of need, not attractive beaches; the worst are just there for the ride and wouldn't be caught dead anywhere unpleasant, like Africa.

Both kinds of science tourist operate at an enormous disadvantage. They're expected to dip into local conditions long enough to suggest or even transfer a scientific solution to a problem without the time to collect useful data. Usually they're more or less totally in the dark because they can't speak the local language and have to rely on local experts whose strongest motivation may be to finagle a trip abroad for themselves.

And, distressingly often, they seem to have no clue about what fools they look. In Belo Horizonte once, dinner out with Brazilian friends was soured when we were joined by an expert from an American government agency, in Brazil to promote his computer model for watershed restoration. Peter was in his forties, in white Bermudas, a yellow tank top, baseball cap and wraparound shades. He had a digital Sony camera in one hand and his twenty-year-old Brazilian fiancée in the other. He talked compulsively. His computer model had worked perfectly "back home." He gave his silent fiancée a squeeze. "Honey, you're gonna *love* it in Denver." Peter had a cackling laugh and flecks of his spittle kept hitting my arm. I twisted and turned, trying to get away from him. "So, what do *you* do, Brian?" he kept saying.

Peter was a science tourist. My image in the funhouse mirror.

SCIENCE ON THE ROAD

I knew there was a problem with my choice of vocation the moment I finally put my scientific credentials to work somewhere that was not convenient, not comfortable, not even remotely close to home. I went to the Philippines. Under my belt, a degree and a few publications. I had managed to find funds to help with a marine fish-culture project at a remote research laboratory in a place I'd never heard of, on the island of Panay. There was in fact a limitless number of things I didn't know, beginning with how to get from the Manila airport to a downtown hotel (stupidly, I took an unmarked taxi). But the next day I managed to get on a dodgy-looking Boeing 737 at the domestic airport, a place where the operators of concessions sat with their heads buried in their arms and passengers were advised to "surrender all firearms and bolo knives." A small boy wandered about spraying the baseboards — and a ten-seater shrine to the Virgin — with insecticide.

I muddled my way through and, two hours later, found myself staring out a cracked and cloudy plastic window as the plane inched toward the terminal at Iloilo City. The first human being I saw was hunched on that concrete griddle, wearing a wide straw hat with a kind of mantle that fell over his shoulders, cross-legged and tapping at a crack in the surface. The plane trundled close to him and lurched to a stop; he didn't look up. We walked down a swaying aluminum staircase into a blast of saturated air, and the man raised his little hammer and brought it down on the runway one more time before I was swept inside with the others. Seeing that man frying like an egg in a wasteland of concrete was like walking along a familiar road and being blindsided by a truck. It was my first trip overseas.

I was met by the chief of the research station and six of his deputies, but this turned out to be coincidental; he was leaving on the next flight and seemed to need this many people to see him off. But someone had known enough about my arrival to send a car containing three men I never saw again. They took me for lunch at a sweltering hole in the wall in Iloilo City, giggling while I did my best to eat whatever was in the bowls in front of me. Do you know what *this* is? Can you *imagine* what you just ate? Much of it was liquid and the angry purple of cooked blood. When they left me in front of the guest house at the biological station in Tigbauan, I was glad to see them go.

The trouble was, nobody came to take their place. It was Friday evening, the compound was emptying, and although I could see people moving between the buildings I was too shocked and shy to approach them. I went inside, lay down on one of the beds and passed the first sleepless night of my life: the famous author of *Induced Breeding in Tropical Fish Culture,* curled up and whimpering in the dark. It was hot and I had never been hot before. There was a gecko stuck to my wall who went *uck uck uck* every five minutes, and I had never seen a gecko. Sirens sounded from time to time. There were military checkpoints right outside the station because the New People's Army — the military arm of the Communist Party that had a large bone to pick with then-dictator Ferdinand Marcos — were in the hills behind the compound. I had never seen a guerilla.

The place was deserted all weekend. The laboratories and experimental tanks that fronted the ocean held not a human soul, and I drifted through them like a pale ghost, peering into bubbling seawater incubators and lifting the lids on circular swimming-pool-sized tanks to glimpse the shadows of metre-long milkfish and sea bass cruising in endless circles. I sat on a pile of rotting nets and watched children pushing triangular sleds through the surf, collecting larvae to sell to fish farmers.

I figured out where the cafeteria was and sat there with my bowl of rice and warmed-over chicken, and in desperation I grabbed a blank notebook I had brought along for recording scientific data and began to take notes: the food, the sensation of watching a jet pass overhead, anything to keep from going crazy. They were not notes about biology, at least not about fish. I scribbled down that the armies of toads that materialized at night "sat like little desiccated Buddhas," which wasn't bad, considering. And that I tried eating *balut,* a hard-boiled twenty-one-day duck embryo; that you could feel the feathers; that people preferred to eat them in the dark.

There was one other person there, a German consultant who intimidated me even more than the empty cafeteria. Volker Storch looked to be in his forties, with a world-weariness that made me deeply envious. Sometimes I would catch him pushing his chair away from the table as I came in, sloping off toward the outdoor terrace with a cigarette stuck to his lower lip. Other times he would appear in the doorway of the cafeteria as I was finishing up, hunched and wraithlike, moving along the counter with his plastic tray and shaking his head. There were hardly any choices and most of them involved warmed-over rice. He spoke only once, and that was the time dinner featured shrimp. They were cooked beyond recognition, curled like wizened fingers in a shiny puddle of oil. The sepulchral German set his plate down opposite mine and chewed his way methodically through a soggy pile of them before looking up. He put a cigarette in his mouth, lit it, blew a stream of Marlboro smoke past my shoulder.

"Zey are trying," he said, "to turn protein into *diamunt."* Then he picked up his empty plate and left. The cook took pity on me and poured me a very large glass of Tanduay rum mixed with *tuba,* a liquor made from fermented coconut. I returned to my guest house and my gecko, weaving.

My isolation ended the next day. Contacts arrived, introductions and fusses were made, hands were shaken. And then the unfortunate news: the fish I had come to work on, the big, silver, mature milkfish for which I had planned an ambitious several weeks of experimentation, would not be making an appearance. Something about a dispute with fishermen contracted to catch and deliver them to the station. I never was sure, but the meaning was clear: I would have to wait, or work on something else. A real biologist would have settled in, rethought his plan, adapted, taken advantage of an abundance of scientific questions few even get to dream about. Instead, I ran.

I hooked up with a young woman who had been a graduate

student in Vancouver, convinced her to act as a tour guide, and within a day I had made excuses, packed my bag and the notebook that was starting to get filled, and returned to the capital. I stayed in a hotel, Zonie stayed with her parents, and together we spent three days exploring the city. I interviewed a trio of young nuns on the seawall in Manila Bay; I saw a colonial pipe organ as magnificent as any in Europe; I clutched my bag and followed Zonie through streets whose every block contained more humanity than I had seen in my life. I threw money at throngs of youngsters until Zonie told me to stop. I learned how to bribe a policeman. After a couple of hours my finger-nails were black from the pollution and my hand stuck going into a pocket for limp bills. From the backseat of a car I stared down the dirt grid of a slum outside the city, horrified, and I snuck a photograph of cardboard and corrugated metal and angry faces.

And I wrote it all down. Of Manila, my notes said: "Imagine a singer, alone, studying a new piece of music, tapping his fingers, perhaps trying out the easier passages *sotto voce:* That's Victoria. Now imagine the sextet from *Lucia di Lammermoor,* in the same room, lustily: that could be Paris or Rome. Now pack the room with fifty tone-deaf madmen, each shouting at the top of his lungs: Manila." And about the public toilets: "Which is worse to sit on — urine or spittle?" The heat was so bad "even nudity wouldn't help." I thought, at the time, that taking notes like these were the actions of a coward, but I was wrong. They were simply not the actions of a biologist.

I WAS A BIOPIRATE IN THAILAND

In 1985, three years after the debacle in the Philippines, I walked through U.S. customs in Los Angeles carrying a duffel bag stuffed with leaves. The leaves were still fresh. They came from a tree that grows in southern Thailand. I'd collected them the week before, as part of a research project aimed at identifying new botanical pest-control agents. More accurately, I'd taken notes and scampered around picking up leaves and bits of branch while Andrew, the local botanist contracted to locate the plant species on my list of candidates, thrashed around in the canopy with his six-metre pruning stick, chopping off bits of tree for me to test back in my lab in Victoria. That's still how it's done: grind up some leaves, extract them in alcohol, add a few drops to whatever you want to affect — cancer cells, bacteria, in my case fish — and see what happens. It's called a bioassay. Back then, Andrew the local expert was just a character. Today, he would be a biopirate.

Research is like fashion; its hemlines go up and down. Scientists use the tried and true scientific method, but the way they actually apply it obeys unwritten rules that keep getting revised, year after year. The changes are usually gradual, unless you look back far enough. Then they're enormous. For naturalists, the golden age must surely have been in the mid-nineteenth century, when millions of species were still awaiting description, when a generalist like Sir Richard Burton could fill his notebooks with observations about rocks and plants and people and customs and all of it would be new.

It took hundreds of years for the world of the old-style naturalist to shrivel to the level of the single gene, where so much of the work is being done now. Even as recently as the 1980s, the world was still pretty much wide open to biological exploration. That was before the International Convention on Biological Diversity, an event that rewrote the rules on research on biodiversity in developing countries. When all those delegates converged at the Earth Summit in Rio de Janeiro in 1992, they thought they were going to change the way nations conserved and shared the world's natural riches — and they did. What they probably didn't realize was how much they would change life for Andrew.

Andrew had a past, only a fragment of which I ever learned. He lived in Thailand, spoke fluent Thai, had a Thai wife named Pen who came with us on the collecting trips and sat, cross-legged, on the ground beside him when he had descended from the tree canopy. Together, she and Andrew placed samples of every leaf into an old-fashioned wooden plant press, after trimming the stems with a fat Swiss Army knife. I guess that made her a biopirate too. Andrew had been an American once, had worked in Singapore as curator of a herbarium, had migrated farther south for undisclosed reasons and was now available for hire to anyone interested in the fantastic flora of Southeast Asia.

Andrew seemed to know the name and location of every tree, every creeper, every miniature pink bromeliad captured in the crook of a vine ten metres off the ground. Somewhere along the line he must have contracted malaria, because when I worked with him he trembled and sweated like a racehorse, his short-sleeved shirt not just stained with perspiration but sodden, not a dry spot left. He wore a dirty bandanna around his head and every half-hour or so he would slip it off and wring it out, the sweat plopping into the dusty soil. His glasses kept slipping to the end of his nose.

And he never stopped talking. Even with his body wedged halfway up a tree and the pruning pole thrashing around in its branches, he kept up a steady stream of chatter and grunts, his voice indistinct but his location betrayed by the furious shaking of a branch or a shower of foliage. Sometimes he puffed and snorted, as though some kind of big animal were trapped in the branches.

He kept it up on the ground too. One time we sat in a tiny restaurant in Hat Yai, a messy town near the Malaysian border, just me, Andrew and the silent Pen. Hat Yai was a place greatly appreciated by the busloads of Malaysian men who by simply crossing the border could escape the strictures of Muslim society and dive into the wide-open and highly affordable world of Thai sex. Hat Yai was a place where the bedside table in your hotel had a brochure rack stuffed with flyers for a half-dozen local girls. If some of the faces looked familiar it was because you'd already run into them in the lobby. Chances were that they would have smiled too — not the cartoon leer of the fallen woman but a big, friendly putting-you-at-your-ease grin that had to be one of the secrets of the phenomenal success of Thai prostitution. One scientist I travelled with in Thailand confessed himself unable to resist. Listening to him, it almost seemed as though the sex had been unplanned, an afterthought.

"I simply went into this little shop to get a hangnail clipped." He held up the guilty digit. We were eating breakfast in the hotel. "Then they wanted to give me a massage."

"They?" I wondered where he had disappeared to the night before.

"Well, at first there was only one. But, Brian, they were so *friendly.*" He looked almost apologetic. "Do you know, they came back to my room the same night?"

How many were we talking about here? My colleague wasn't a young man.

"And they keep phoning me! So friendly! So inexpensive! Brian, you really *must.*" He suddenly put down his fork. "Would you mind awfully picking up the bill? I really can't face them just now." He dropped his napkin on the table and darted into the street. In the door to the dining room I saw several young women scanning the room. He was right, they *did* look friendly.

Andrew's experience had been a little different.

"These guys, these Malaysians and the rest, they pay for it," he told me. He slurped his soup. I guessed "the rest" would have included my colleague. Andrew was still wearing the sopping head-band. "My friends and I, there was a bunch of us expats, we never had to pay for it. We used to meet every Wednesday night in the same bar, up in Songkhla. A half-dozen beers, and the girls came to us." He reached for his bottle of Singha and I wondered if he wouldn't have been better to pour it over his head. He appeared to be steaming. I took another swallow of the soup: mussels and lemongrass, fiery beyond belief. All you could do was follow each mouthful with rice and try not to cry. I was sweating too.

"You didn't worry about, you know, disease and stuff?"

"You want to know the secret? I'll tell you. Soon as you're finished, you just pour a big glass of that cheap whisky they have here and dip it in, swirl it around a little. Easy."

"Very scientific," I said, wondering what happened to the Golden Cat after that. Maybe he tossed it back like a chaser. "I'll remember that." I looked at Pen, bent over her bowl of soup. It was almost empty and she wasn't sweating a bit. She looked up and smiled radiantly at me.

"Doesn't speak a word," said Andrew. "It was a regular club we had up there, even had a name for it. The LBFMPBR Club."

"Oh, yes," I said. Andrew's red-tinged eyes had a fond, faraway look.

"Little brown fucking machines," he said, draining the rest of his beer. "Powered by rice."

Maybe Andrew's dead now, a victim of malaria or AIDS or gravity. Maybe his silent smiling wife was nursing unspeakable grievances and leaned over one night while he sweated and muttered in his sleep, pulled out the Swiss Army knife and lopped off his dick; this happens in Thailand. If he's alive, he's certainly out of business, because the collection of biological samples from developing countries has been all but shut down. Permission, if it's attainable at all, now theoretically has to come from the local community, an entity that the Convention on Biological Diversity has enshrined as its number-one client and awarded extraordinary powers. Of course, like all powers awarded through international treaty, these ones are non-binding and almost entirely illusory. But no matter, it's still become virtually impossible to do the kind of Wild West field research I hired Andrew to help me with. The fashion has changed.

How science works (and why it can be so boring)

Hollywood loves scientists because they're always proven right. That means the script can safely subject them to all kinds of abuse ("I'm sorry, doctor, I don't have time for your theories now, my president needs me") before the five-star generals have to extend a callused hand and eat crow: "Well, dammit, Doc, I was just plumb *wrong*. There really *was* a tsunami (or a major earthquake or a triceratops) headed right for the Empire State Building."

Decision makers love the phrase "science-based." They're addicted to it. It's comforting, and I know it works, because for nearly a decade I ran a conservation NGO and I used it all the time. I would smile at the guys with the money, the suits, and say, "We're science-based," and they would nod agreeably. It was just what they wanted to hear. Actor Bill Murray caught it perfectly in *Ghostbusters*. "Back off, man," he said. "I'm a scientist."

You would think — hope — that conservation is science-based, but it isn't always. Science may even be the least important thing in many conservation campaigns, because campaigns are all about changing people's behaviour, and to do that you need to deliver a compelling and unambiguous message that anyone can understand — not only the people whose minds you want to change, but also your funders. Science needs time to develop a clear picture, a requirement that's antithetical to the realities of operating an environmental organization. If a conservation group has to choose between spending money on doing its scientific homework or hiring a good PR firm to stay afloat, that's not much of a choice. And tried and true images are hard to abandon even when the science doesn't back them up. For example, lots of consumers still believe what so-called science-based organizations have told them about the "chemical additives" that make farmed salmon's flesh the right shade of orangy-red. But the additive is only astaxanthin, a natural antioxidant that's been colouring wild salmon for as long as they've been on this planet.

A decision with science behind it seems, if not unassailable, at least defensible. So science-based conservation should be a good thing. The trouble is that, to do it on any meaningful scale, you have to dumb the message down to the point where it isn't science any more. I doubt many decision makers understand how science works; worse, there seem to be more and more scientists who've forgotten their training. For all of those people, and for anybody else who's curious about what "the science" really means, here's a primer in the form of some personal history.

The scientific method is elegant and simple; the problem is knowing its limits. First, you ask a question, say, "I wonder how those little luminescent shrimp turn their lights off when you blast them with a flash gun." This particular question seems trivial, even idiotic. That doesn't matter. All that really matters in a scientific investigation is that you test your question according to certain rules and that you 'fess up with the answer. The Ph.D. I used to wangle my trip to the Philippines I got for answering this particular question, idiotic or not. The little luminescent shrimp were quite beautiful, so transparent you could see their tiny hearts contracting. They really did produce light, and they really did turn their lights off when you flashed them. How did I know this was true? Because some other scientists had demonstrated it.

So you dream up a hypothesis, which in my case was, "Well, they probably have nerves going to those light organs and the nerves fire and the light turns off." Seems reasonable enough, but more importantly it's testable. All I had to do was get my hands on enough of these shrimp to work with, figure out a way to get a microelectrode onto the nerve going to one of the light organs, wait till they lit up and then *zap!* with the flash gun and watch the wiggly pens. In the downtime I would make reams of electron micrographs of the nerve and the light organ so that, when the great day came and the nerve sent out its little spike of electrical energy at the predicted moment, I'd have the pictures to back me up. After that, my thesis committee, a half-dozen academics with thriving research programs and tenure, would grill me for a few hours and hand over the keys to the executive washroom. Go forth, and publish.

But I'm getting ahead of myself. For all these wonderful, career-launching things to happen, you have to prove or disprove your hypothesis. To do this, you run it through the reality check of experiment, and it's in this process that the beauty of science becomes apparent. Because of course nothing works the way it's supposed to. Science is, or should be, all about *admitting what really happened;* forgetting this crucial point is where decision makers go wrong.

In my case it took two years to find out how wrong my idea was: the damn shrimp, which are glasslike, less than two centimetres long, and live fifty metres down, could only be caught at night in the middle of a certain inlet; the microelectrodes had to be hand-pulled from glass tubing the diameter of a thread (a process I never really mastered, littering my lab with the filamentous rejects); the electronics of recording nerve impulses stymied me; actually getting a transparent electrode onto a transparent nerve inside a transparent animal was nearly impossible; and my animals, which I tried to keep alive in polyethylene four-litre jars, kept dying.

"Brian," a muffled voice would come from outside. "You've been in there three hours. Everything all right?" That was a question I had trouble answering.

And of course, when after countless hours in a dark room hunched over a microscope I finally got everything set up and flashed the poor creatures, nothing happened. Flat line. The animals' lights dutifully blinked off, but the nerve that was clearly going to them wasn't doing a thing. This is the kind of scene a cost-cutting politician lives for: some idiot locked in a darkened laboratory hunched over an innocent shrimp he's been keeping alive at gov-

ernment expense for the sole purpose of sticking invisible glass electrodes into it, all in the service of a harebrained hypothesis that's
palpably wrong. I did tend to wonder myself.

But that's the best part about science: when you know you're
wrong and you have to dig yourself out, save face and publish something that's *not* wrong. The beauty of the scientific method, and the
part most people don't understand, is that it forces you to abandon
a hypothesis if it doesn't fit the awkward facts. In desperation, I took
my supervisor's offer to spend a summer in France, where one of his
colleagues worked on a cousin of my doomed shrimp in research
stations on the coast of Brittany and in the town of Villefranche,
near Nice. The French shrimp were bigger; maybe they'd be easier
to work with.

Jean-Marie Bassot was tall and morose with a crumpled,
Thurberian face. He was all shrugs and charm. I instinctively felt
Jean-Marie would put me right — except that it rained and blew for
a solid month in Roscoff, the coastal town that was my first stop. So
instead of going to sea to collect shrimp to stick more needles into,
I took a lot of long walks to the end of the breakwater (it smelled
strongly of piss) and learned how to drink Algerian wine cut with
water. In Villefranche, where I spent all of August, I learned that all
of France takes its holidays in August, including the people who were
supposed to help maintain my experimental animals. The few I could
catch turned up their Gallic tails and died instantly. In Villefranche, I
wandered around buying baguettes and ham to sneak into my room
for dinner because I couldn't afford to eat out, and watched the glistening white gin-palace yachts come and go in the harbour. The
research station was originally the town jail; you could still see the
attachments for the leg irons, and the locals swam directly under my
window. One day I watched a little girl lower her bathing suit and
relieve herself while her brothers and sisters splashed around her.
They played on, surrounded by a flotilla of tiny turds.

Those turds were just about the only sea creatures I saw that
summer, but science marched on, as it had to if I was going to finish
my degree. My first hypothesis was clearly wrong; further testing of
it in France wasn't working out. I was heading back to Canada soon,
so what could I test next? I needed a new hypothesis. On one of my

long walks up the winding road above the harbour I got what's usually described as an epiphany: the whole lights-out thing had nothing to do with nerves. It was an artifact of somebody playing around with flash guns. The phenomenon was a trivial dead end — but I could use the scientific method to prove it! They *had* to give me my Ph.D.; some of the guys on my committee had made entire careers out of sillier subjects.

As soon as I got back to Canada I fired up my outboard motor and headed into the inlet in the middle of the night (the shrimp came up closer to the surface then), trawled up some specimens, and instead of going through all the rigmarole of electrodes and microscopes I just popped a dozen or so in a blender with some seawater. In two seconds I had a glowing cocktail as the light-producing chemicals in the shrimp's bodies mixed and set off their reaction. The liquid was a cool blue-green, the test tube glowing in the dark lab like some kind of futuristic dildo. I held it next to a flash gun and *zap!*

Out it went, instantly. I sat there in the darkness, not sure whether to laugh or to cry. Obviously (or obviously enough for the march of science), the whole thing was just a chemical reaction, a bleaching that had nothing to do with nerve impulses or any intact animal at all. Ten minutes later the light faltered back on, until the test tube was once again glowing in the dark, just a soup of shrimp and seawater. I zapped it again; out it went again.

This was science. The fact that it was applied to something trivial was irrelevant. I wrote it up as a scientific paper, and I have no doubt that other biologists interested in luminescent organisms read it and checked it out in their own laboratories. Whether my results were confirmed or refuted I have no idea because I rapidly lost interest in the subject, but that doesn't matter either, because science, the best science, works just this way: someone publishes some results, which suggest further experiments to someone else, and the whole enterprise lurches from hypothesis to hypothesis until eventually a true picture of the way nature works emerges. True — not morally right or wrong, just true.

The part that confuses people is the utility of it all; they think there should be some usefulness. Maybe there's some light-sensitive chemical in that shrimp's system that has inestimable value for humankind, a cure for cancer or baldness or impotence. But usefulness isn't built into the scientific method. Science doesn't care.

Or perhaps I should say, *good* science doesn't. The best science is the result of disinterested exploration, a kind of muddling around in nature while keeping good records. The great Darwinist Thomas Henry Huxley, an almost exact contemporary of Richard Burton, was an impassioned defender of the scientific method and surely its best epigrammatist. It was Huxley who wrote about "the slaying of a beautiful hypothesis by an ugly fact" (luminous shrimp, anyone?) and offered the stern advice to all scientists: "God give me the courage to face a fact though it slay me." The scientific method only really works when it's disinterested, when scientists can follow their noses along the hypothesis-testing chain, wherever it leads. If someone has already decided what the goal should be, and the response of nature to all that poking and pin-sticking turns out to be completely different and unexpected, real science has to accept that. Real science doesn't care what the desired result was; unfortunately, funders often do. Again, Huxley put it best: "Science adopts suicide when it adopts a creed."

Science, the actual practice of it, is surely one of humankind's greater inventions, because it manages to combine the sense of wonder with a second (but definitely not secondary) requirement: the need for verification. This second part is unforgiving. As a scientist you can dream, but you have to prove your dreams are right, or at least try. Science is not, at its core, creative. You can have a perfectly good scientific career with the balance tipped way over in favour of the verification part, with just enough dreaming and speculation to provide hypothetical fodder.

But you can't do the reverse. Experimentation, collection of facts, publication of the results: that is your real business. If you happen to find such things boring, you're in trouble. I noticed this depressing fact at the first scientific conference I attended, and have been noticing it ever since: scientists, when they're at the business of formally presenting their science, are boring. That first meeting was in Hilton Head Island, South Carolina, and after half a day of papers on the pH of fish gonads I was slamming along at twenty knots in a rented Boston Whaler, exploring the mangrove forests. I still feel guilty about this.

Scientists cannot afford to be lyrical. Their hypotheses, it's true, can be fanciful, even dead-on brilliant, but there's no escaping the cornerstone of science, which is proof. The day-to-day business of the scientist has inevitably to dance attendance on facts. And to get facts you have to ask simple questions. This leads to the problem of endless reductionism. A scientist can wonder, as can an artist, about nature's mysteries but, in order to communicate with other scientists, he has to confine himself not just to one mystery, but to some sub-subset of that mystery, biting off just enough to ensure irrefutable answers. Most people don't realize just how confining this process is, so I'll make up an example.

Let's imagine you're fascinated by life in the depths of the sea. You can dream about it all you like, but if you're going to study it as a scientist you need to specialize. What should you study? Let's say you've always been fascinated by little things, the stuff that only reveals its beauty under a microscope. But the ocean is huge; micro-organisms abound on the seashore, in mid-waters, in abyssal zones where light never penetrates. Abyssal zones sound mysterious and compelling, let's go there. But there isn't much money just to dig around at random for something interesting at the bottom of the sea. So you have to go after something more specific, like the micro-organisms that live near hot vents miles below the surface, weird things that can survive in water percolating up from the earth's core. Bacteria live there, some pretty odd ones; maybe you'd better study them. Bacteria might be worth something. So far that sounds pretty interesting, with a good chance at funding.

But what aspect of the bacteria's life can you study? How they adapt? The strange chemicals they produce that might represent a cure for cancer? Either of those means you're going to have to be a biochemist, comfortable around centrifuges and chromatographs and test tubes. More narrowing down, because remember, you have to be able to ask science an *answerable* question. And so your field of specialization gets whittled down until, at the end of the day, when someone at a party asks you what you do, your answer will now go something like this: "I study the sodium reductase pathway in the

Kowznofski reaction that's believed to be involved in a novel mito-
chondrial synthetic chain in thermally adept marine bacteria."
Suddenly, a fascination with the sea looks a lot less romantic.

Of course, the fact that the process of science is fundamentally
boring doesn't mean its practitioners have to be. There are plenty of
interesting scientists, even fish scientists, although the way their
minds work sometimes means that a conversation goes down pathways whose logic takes some time to be revealed. I sat in my
backyard once with Blair Holtby, a Canadian fisheries biologist, and
he couldn't seem to get his mind off my fig tree.

"Looks like you're getting a second crop of figs this year," he said.
Blair is roundish and bearded. I picture him reading poetry in a
smoky room. He slipped his sandals off and wiggled his bare toes in
the sun.

"I guess," I said. "They never really do much, though." We both
gazed at figgy nubbins the size of marbles. It was August already, the
days were getting lamentably shorter, it was spawning time for
salmon. "But definitely bigger than last year." I couldn't see where
this was going.

"That's it, bigger than *last* year. My dogwoods bloomed twice
this year. Climate change. The plants know it. Don't go swimming
up in Hecate Strait."

"Excuse me?"

"Up by the Queen Charlottes. There's *mola mola* and great whites
up there right now. Temperatures in the teens."

At least we were back to fish. Half an hour ago we had spent fifteen minutes on the role of certain cat foods in kidney infections,
which had led to a long story about the gestation period of guinea
pigs and how hard it was to predict ovulation and how those two
facts had led to a population explosion in the Holtby household
menagerie. To Blair, all these things were biology, logically connected: animal physiology led to animal behaviour which
contrasted with plant reproduction which was influenced by climate
change which made it dangerous to swim in waters that were normally too cold for great white sharks. Blair was a perfect scientist:
he saw patterns everywhere. He was probably the kind of person
who could look at one of those computer generated nonsense

pictures and "see" the hidden pattern in seconds. I would stare for hours and see nothing but, eventually, spots.

"We were talking about this because of the cat food," Blair said. He wasn't lost at all. "That's why management is so confused."

"Management?" I sometimes feel stupid when talking to Blair.

"The Fraser sockeye fishery. The reason it was closed this summer is because it's just too confusing. Nothing to do with this group or that group, political pressure, whatever. Ocean conditions, river conditions, they're all extremely unusual, so management's handcuffed. They don't know how many fish are coming back, and when, and what the water temperature's going to be in the rivers. The salmon are just acting weird." He waved in the direction of my garden. "Like your fig tree."

For a time, scientists could be true explorers — and explorers, scientists. Explorers like Richard Burton in fact contributed quite legitimately to the scientific knowledge of their time, because there was so much still to be discovered, including plants, animals, rock formations, even new human societies, that anybody who kept decent records could publish his findings and make a real contribution to knowledge. Maybe the last romantic journey made by an explorer of the old school in Brazil was the one made by Colonel Percy Harrison Fawcett in 1925. Fawcett was fifty-seven, nearing the end of his career, when he set out with his son and a young friend to find, once and for all, what he would only refer to — even to his backers — as the "Lost City of Z," the ruins of an ancient city he was convinced lay in the heart of the Brazilian Amazon forest. Fawcett is rumoured to have been the inspiration for the character Indiana Jones; his monomania and secretiveness reflect a time when there were still firsts to be won. As he put it, recalling how the Norwegian explorer Roald Amundsen beat the British Robert Scott to the South Pole by a matter of weeks, "There can be nothing so bitter to a pioneer as to find the crown of his work anticipated." Fawcett vanished, fading into the forest like the era he represented.

SCIENCE THE WHORE

Now, most of the discoveries that could have been told directly to the man in the street have all been made. The wonder of science has long been replaced by the blind, trusting belief that it will keep churning out discoveries that will make our lives easier. These days, society wants *results* — cure cancer, build a better fuel cell, bail us out of climate change — and this strains the fundamental follow-your-nose character of science. It's been a long time since societies could justify paying scientists to do nothing but the "pure" research that flourished in the eighteenth and nineteenth centuries and was fed by a patronage system. Today we have "applied science," which is used to produce the products and services societies demand.

When you don't really understand how science works, but feel certain it can do good things for you, you may also overestimate the individual scientist's importance. People do this all the time, when they intrinsically trust someone with "Ph.D." after their name, and bureaucrats are especially prone to this mistake. I once received an unannounced visit from two men in suits. One of them handed me a card. There was a name and a phone number; that was all. I remember his first name was Randall.

"We're from CSIS," he said. He tapped the front of his card. "You can call that number to verify." CSIS was the Canadian Security Intelligence Service. Our CIA.

"That's okay," I said.

"You're doing a lot of travelling," Randall said. "Something about technologies for fish conservation."

That rattled me. I wanted to say, how did you *know* that? Instead, I said, "Please sit down," except that there was nowhere for them to sit. My lab was an alcove, borrowed from the university, about the size of a decent hanging closet. The three of us were already uncomfortably close. Randall used cheap cologne. I got up off the only seat in the room, my stool in front of the microscope.

"That's okay," said Randall. "We won't keep you long, Dr. Harvey." Nobody called me Dr. Harvey. "I'll get right to the point. CSIS has concerns about the theft of certain technologies by foreign nationals. This can happen especially easily during travel."

What planet were these guys from? My technology was a Rube Goldberg adaptation of something the cattle industry had being doing for thirty years. It allowed the lucky user to freeze fish sperm. I practically had to beg people to hang around while I demonstrated it. Its value on the science black market was less than zero.

"I don't really think . . ." Randall held up a hand.

"Don't think, Dr. Harvey. Just keep your eyes open, that's all we ask."

He stuck out his hand, they both did, and then they were gone. They weren't from another planet, they were just from government, from a department especially removed from reality. But the mistake they were making with me was fundamentally no different from the one that's being made every day about what science can do and what it can't.

The problem is especially bad when science is used to provide a service to society, like natural resource management. If the numbers go against what the managers want or, worse still, there isn't enough money even to get good numbers, "science-based" becomes a farce. This is the reason so many fisheries are in trouble today. Either there's not enough money to do the science needed to fix the problem, or (more likely) there's been enough solid science for years to predict what was going to happen, but decision makers simply ignored the warnings. Usually the science comes from their own scientists, too. The northern cod are always the best example: science knew what was going on years before the cod finally crashed, but the message was something the decision makers just didn't want to hear.

Science is no longer the dewy-eyed virgin it was when dreamers and collectors like Burton and Fawcett could roam, limited only by their ability to cajole funders and endure ticks and mosquitoes. Science is whorish now, rouged and loose, a service industry for social engineering and technology. If dreams are what you need, you're unlikely to find them in science now; if you do, you'll spend most of your time defending them from a fate worse than death.

Science and politics is an especially unpleasant mix. A friend in FAO groaned when I asked him why he was travelling to Indonesia in early 2005. It was not long after the tsunami. "Don't ask," he said. "Please." And then, of course, he told me anyway, in the way that people feel compelled to spill their feelings about the idiocies they get dragged into perpetrating. Confession makes it better.

"To rebuild all the boats destroyed in the tsunami," he said. I could almost see him hanging his head in his office in Rome. Maybe he was pounding it against the desk. "I know, I know, we've been saying for decades the fisheries are overextended. Trying to come up with ways to reduce fishing capacity, retrain fishermen, all that stuff. Spent millions on it. So along comes a tsunami and in ten minutes it removes a big chunk of that over-capacity and what do we do? We spend more money to help rebuild it! It's nuts." FAO and other agencies are doing it, of course, because the world wants to see something done, *now*, and so do the fishermen. In the end it's easier just to build more boats. The heck with the science.

CHAPTER 8

SEX WITH FISH

The art of spunking

Trust me, this won't hurt a bit: getting ready to spunk a farmed Atlantic salmon, British Columbia.

I GOT INTO FISHERIES THROUGH THE BACK DOOR: hatcheries. It was hatcheries that first brought me to the São Francisco River. Hatcheries are all about the eternal truths: sex and survival. In hatcheries, sex is mechanized, an out-of-body experience for the fish. If you work in a hatchery you spend a lot of time with your hands full of eggs and sperm. I was an expert on sperm. For a time, I promoted gene banks as a way of conserving fish stocks, and I handled gallons of the stuff because that was where the genes were. It was applied science with a vengeance.

On the west coast of Canada, salmon technicians favour the Victorian vulgarism "to spunk." The providers (the males) are called "bucks," the females "does." Someone would toss a pair of stained rubber overalls into the back of a truck and say, "We'll head on out to the pens, get the bucks sorted and spunk 'em." Women biologists used the same word, although that may just have reflected a selection process that favoured hardy types for whom getting wet and freezing half to death and cursing were matters of honour. I never noted any difference in the skill of male and female spunkers, although many had their own personal style, the same as would a sample of chefs preparing, say, a *roulade* of veal.

Sometimes you spunked in a river, sometimes at a hatchery, sometimes on a fish farm. I worked several times with a guy called Frank, who managed a remote salmon farm on the coast of Vancouver Island. Frank had his own special routine for obtaining top quality sperm from a ripe salmon. Good sperm made good babies, and it was worth a lot of money. Frank's sea cages were reachable only by boat, a few hundred metres offshore from Douglas firs that came right down to the water, their lower branches razor-cut at the tide line and their tops concealed in mist. The pens had metal walkways and automatic solar-powered feeders that spun the high-fat feed onto oily water. Occasionally a bubble of gas from a decaying fish would wobble up and burst, a sign that a diver needed to be sent down again to clean out the vile, disintegrating "morts."

Frank had a hoarse voice, a missing finger joint and a dry sense of humour. "First you gotta just goose 'em a little," he said, slinging a twenty-pound chinook under his arm. He did this all morning; it was an assembly line. I was glad it wasn't my arm, because wrestling a twenty-pound salmon is like trying to subdue a fire hose. It was cold and drizzling and the big fish flexed mightily a few times before Frank had it under control. Their little struggle sounded like a street fight, a flurry of grunts and slaps. Finally he held the fish flat across his stomach, its abdomen twisted so the vent faced out. The head was buried in the crook of one elbow and the tail gripped with a gloved hand. The glove was to preserve the mucous layer which, if stripped off in the struggle, would bring death even sooner than Nature had already scheduled. Not that it mattered to this fish; this one had been captive all its life and would never even see a river.

"Just a little squirt, get all the crap outta there." Frank ran his thumb and forefingers along the fish's silver belly, broad as a plate. A driblet of urine and feces leaped out and fell at my feet. I stood my ground. It doesn't do to shriek and jump out of the way, not if you want to be invited back.

"*Oh* yeah," Frank said, "get that bag up there and let's *spunk* this bad boy, see what he's got." I fumbled in my pocket for a Whirlpak, a plastic envelope with a wire closure at one end. Frank squeezed the fish's midsection again, much harder this time, running the length of the belly and producing a fine forceful stream, pure and white. I had to jump to catch it in my bag. "Hoo, got ourselves a *real* stud here!" Frank ran his fingers along the fish again and again; I darted back and forth and my little bag filled. It was an odd game of catch.

"See now, this is the way you tell when to stop," Frank said.

A sockeye salmon gives it his best shot. Birkenhead River, British Columbia.

"Hear that?" He squeezed hard, near the vent; there was a croaking sound, as though the fish were protesting. "That's how you tell he's empty." I grasped the two ends of the wire closure and whirled the bag so that it sealed automatically.

"I thought you could tell he was empty when nothing more came out," I said.

"Don't forget to write down the number," Frank said and kicked the fish across the deck, where it wriggled, arching its back, until another worker snagged it by the tail and a third man, Carl, stepped forward with a baseball bat. The noise it made sounded like a coconut hitting the road. Blood whirled.

"You look at the eyes," Carl said. "If it sort of stares, it's dead. If it looks up or down, hit that puppy again." He tossed the body into a cart already filled with spent fish and frothing with mucus. So much for the miracle of the salmon life cycle.

I never could feel comfortable around this sperm-sampling, even if it was for my own projects, even if I did it myself. There wasn't much choice, because sperm was the most convenient repository for all that "priceless genetic diversity" I kept nagging people to try to conserve, and I needed sperm to perfect my method for preserving it. Fertilizing eggs in a bucket was the time-honoured way of making fish babies in a hatchery. But it was a squalid business. Most of the time, with salmon at least, the procedure was terminal; the fish was only ever released to spawn naturally if it had been sampled directly from the river, and this didn't happen often. Either way, they were going to die. Mostly it was assembly line carnage, the males squeezed empty, whacked, tossed on the pile to be turned into fertilizer.

The females had an even worse time of it: they were suspended by one person and slit by another, from vent to where their neck would be if fish had necks, using a wicked little device you slipped over your middle finger, like a knuckle duster fitted with a curved, razor-sharp blade. The females fell open like a burst suitcase, a gallon of bright red roe gushing into a waiting bucket. At one farm, the gills were cut and the bodies hung upside down on a long rack, tails spiked and handfuls of paper towel stuffed in the gill cavities. Some sampling or other was the reason; hanging there in the rain they looked like victims of a methodical gangland slaying. I do remember that the person doing the belly-slitting was a woman.

I always felt like an interloper at these spunking sessions. George Plimpton, the writer of *Out of My League* and *Paper Lion*, described with elegance and humour the experience of an amateur accepted, for a brief moment, into the world of the professional. There's a passage in *Paper Lion* that has always stuck with me, where Plimpton is the "last-string quarterback" in the training camp of the 1963 Detroit Lions football team. Throwing the ball — getting the right grip and the perfect release and feeling the ball snap out of your hand spinning fast and hard — always eluded him. He stayed after practice and threw to ballboys and hangers-on and, late at night when the real players were out carousing he threw into the mattress, but he never really got it. Plimpton had his spiral, I had my spunking. I tried to keep my distaste hidden, and whenever I worked with salmon that wasn't difficult, because the fish were so valuable few technicians wanted anybody else touching the fish anyway. But I got very good at holding plastic bags, and learned to give a manly chuckle whenever an unusually forceful spurt of sperm hit me in the chest.

"Some spunker you got there," I would say, careful to let the stuff drip off in its own good time. "A real bad boy."

"Fuckin' A," someone would reply, and I would know my secret was still secure.

Everything else about the procedure was fine — catching wild fish especially, because if you had managed to get permission to take fish from a river it usually meant you were in a beautiful place. Fall colours, a dusting of snow on mountains, the river running hard with the first rains and the torpedo-shapes of spawning salmon lined up just beneath the dimpled surface like a forest of quivering compass needles; you might have to share it with bears but usually you had a shotgun somewhere in the kit. In the big rivers like the Thompson or the Nechako, you might even go out in boats. I was an eager volunteer whenever a jet boat needed to be driven fast across a swollen river, although this was as much self-preservation as anything else. It was the other guy who had to pay the tangle-net rapidly out of a garbage pail into the V of our wake; he stood the best chance of losing a finger or an arm.

But being in a beautiful place didn't mean the process was beautiful. If it's a full-scale, "Alaska Protocol" egg take, you're in for a lot of blood and guts. Usually these wild egg takes are done at a counting fence, a barrier set up across a river, and the salmon are herded and waiting. The only purpose is to get eggs and sperm for the next hatchery generation, as well as some tissue samples to test for disease. Everybody gloves up, everybody sloshes through pails of disinfectant; the team is like a parody of an operating room. Out comes a slug of kidney, like a cocktail sausage. A few millilitres of ovarian fluid get withdrawn into a sterilized vial. Scalpels are flamed with a Bernzomatic torch and a head deftly quartered to get at the otoliths, the tiny ear bones that can be "read" for clues to diet. The ground becomes littered with discarded latex gloves and rolls of paper towels; overhead, tattered alders release their leaves. In the end, fifty spawning salmon are reduced to a row of ice cream buckets and a few coolers full of vials and plastic bags. The bodies are tossed in the river, congealed blood bright in their mouths, and the bucket of disinfectant heaved in afterward, for good measure.

The hatchery itself is usually in the woods somewhere, near the river, a shed with racks of egg incubators and, on one wall, a sink full of dirty coffee cups and a row of hooks where someobody has attempted to impose order with a felt pen: *Larry, Lee, Rookie.* Outside under the trees, a half-dozen blue fibreglass tanks the size of hot tubs, each one full of tiny baby salmon. I saw one such place where the shed door had been clawed through by a bear looking for the sacks of fish food inside. It looked like firemen had gone after it with axes. The same bear had been seen the week earlier, the hatchery manager told me, balanced daintily on the lip of one of the tanks, scooping up baby salmon.

"We lost twenty thousand salmon fry," Lorraine told me. "I turned around and he was looking right at me with those nasty little eyes. I got the fuck out of there."

DOING IT IN THE TROPICS

In the tropics, fish-biologist machismo is alive and well, altered by climate and species and to some extent decorum. The fish themselves, at least the ones I worked with, are often bigger than salmon, affording more opportunities for wrestling. Sometimes the sampling is done directly from the river, but my role on those occasions was usually bag-holder and labeller, for the simple reason that the cast-net used to catch the fish weighs a ton and is damnably awkward to whirl over your head and launch from a perch on a wet rock. You have to grow up doing that. The circular net collapses like a crinoline and has hundreds of eraser-sized lead weights sewn into the hem; just carrying it is hell.

Many of my sperm donors had to be fished out of holding ponds dug in the earth, or from net cages suspended in a lake or river. And although there are many female fish biologists in South America, I have yet to see one wading into a pond full of thrashing four-foot catfish. This seems to be left to the men.

The average aquaculture pond in the tropics is swimming-pool-sized, a rectangle that would fit nicely in an affluent backyard, bull-dozed to a depth of three metres or so and the edges carpeted in spiky short grasses that mean snakes. Shoes are, of course, uncomfortable and they get wet, so no one wears them. I remember one place where the grass was so heavily infiltrated by mimosa that the ground seemed to cringe as you walked on it. If you turned around you could watch your own footprints silently erasing themselves, like a cheesy special effect from a fifties movie about the Invisible Man.

Dozens of ponds dug into the land are connected by a network of raceways and pipes. All that water creates a huge reflective surface, so that a morning, even an hour or so spent standing around the edges or, God forbid, actually inside one of them, is like being an ant on a disco ball. But standing around is inevitable. You have to wait for nets to be delivered on carts towed by tractors; you have to wait for the opaque yellow water to be slowly drained to concentrate the fish, revealing encrustations of whorled snails the size of golf balls along the edges. This draining of the pond, and especially the moment when men began to kick off their flip-flops, was the signal for me to find compelling reasons to fiddle with my camera or remember sud-denly to check the outflow arrangement on a neighbouring pond. A slow circumnavigation, head down and hands clasped behind the back, could absorb critical minutes while other people waded in and seized the net and ranged themselves at one end of the pond before beginning the slow march to the other. If you were lucky there would be distractions, like fresh baby coconuts to pluck like olives, or tent-sized spiderwebs that engulfed entire trees.

"Where's Dr. Brian?" I would hear in the distance, and I would keep walking.

Because these ponds are not nice places. You enter them on your backside, slithering down the sloping edge. The water is blood-warm, up to your chest, and your bare feet sink into an ooze of decaying feed pellets and fish shit and snails. Any cut on your body lower than your chin gets automatically inoculated with bacteria the *gringo* immune system has never met. Once the crowding march begins, you shuffle forward clutching the wet cord at the top of the net and wishing you had remembered to take off your rings, and as the big fish sense what's happening they begin to reveal themselves

Evoy Zaniboni hoists a farmed surubim *catfish.*

as sudden menacing humps in the water just ahead, or a glancing blow to the knee.

One time, watching one of these sampling sessions, the fellow next to me turned and said, "When I go swimming, I can never shake the feeling that there's something just underneath me, you know?" He shuddered. "Like when you were a kid, swimming in the lake. I remember hanging onto the side of a dock and praying something wouldn't suddenly come up and bite me." He waved at the pond, where a ragged line of men was closing off the last avenues of escape for fifty crazed catfish. "You wouldn't catch me dead in there."

As the fish are corralled the net closes into a tight, boiling circle. To breed these animals, which is normally why the whole crowding routine is done, someone has to reach into the melee and begin pulling fish out bodily, slinging each one under an arm, applying just enough pressure to the belly to expel a little sperm or a dribble of eggs or, if the fish turns out not to be ripe yet, nothing more than urine and feces. The ripe ones get handed back up the slippery bank to somebody waiting with open arms on the edge — another signal for me to consult my notebook or squint furiously into my camera.

Usually the job of obtaining a sperm sample from one of these tropical species was assigned to one of the men who maintained the ponds, and they had their own routines, the same as the people who worked on the salmon hatcheries. Their fingers moved surely along the flanks of the fish, which were often subdued by placing a damp cloth across their eyes, as though they were revellers suffering from a particularly bad hangover. One man did a dozen fish while I watched, and where Frank had a joint missing this man's little finger made up for it with an extra digit sprouting from the last joint. When he milked the fish it wobbled like a grape.

There was one curious difference between the salmon and the big tropical migratory fish, and that was sperm volume. Wherever I went, to rivers and ponds in Colombia, Venezuela, Brazil — countries where machismo is alive and flourishing — I never saw the equivalent of the "spunking session" so common in Canada. In Brazil a man will check his genitals reflexively, in public, a quick clutch and jostle through the trousers as though to reassure himself that the source of his power is alive and well in there. Armani suit or beach shorts, everybody does it. But the male fish just don't measure up. The females are prodigious egg-layers: a good squeeze of a gravid fish produces a thick stream of tiny ova — millions of them. But production in the sperm department is puny.

Once, by the side of a remote river in the state of Minas Gerais, we clustered around a spotted catfish easily four feet long, while biologists anxiously dried its pallid abdomen and awaited the verdict. What came out of him was a few measly drops, like the dregs of a shagged-out porno stud. It's the result of an anatomy that makes it all but impossible to get one's fingers on the right places, plus a fundamental difference in fish reproductive behaviour. Salmon spawn in fast-flowing water that dilutes the sperm as quickly as it's fired over the eggs; the big Brazilian catfish are smarter, moving into quieter lagoons and side channels where the dilution effect is less. They don't need the money shot.

When those few precious drops came, a Brazilian biologist

stepped forward with a small test tube and deftly scooped the sample up. She wore tight black stretch jeans and high-heeled boots and she held the sample up and winked, as though to say, is this the best you can do?

THE LAST PLACE I EXPECTED TO SEE AN ORYX

Villavicencio, Colombia, 1995. Seven sobering hours by car through the Andes from Bogotá, and I was in love and, even though I didn't know it, about to meet my first *narcotráfico*. And all because of hatcheries.

Villavicencio is a noisy, haphazard agricultural town on the edge of the *llano,* the plain that stretches endlessly east once you're through the mountains. It's cattle country, served mainly by the winding road from the capital that's clogged with a never-ending train of trucks grinding along in a cloud of diesel exhaust. Shrines to highway casualties are everywhere, bleached rosaries and flowers desiccated by the sun. On the way from Bogotá, we passed an over-turned car and a man wandering on the shoulder, blood covering his face. Our driver shrugged: the fellow was still walking, wasn't he? Stopping could complicate things. This was country where bandits leaned into your open window and blew smoke from a narcoleptic plant into your face, or so I was told. Whether the story was true didn't matter; like all such stories it served its purpose of ratcheting the paranoia up just enough to keep you on your toes. Never a bad thing in Colombia.

Which explains falling in love with Maria Lourdes. If you're in a place where visitors are told to stay in their hotels in the evening, where jogging beyond a few city blocks is an invitation to a kidnapping, what else are you going to do when someone introduces you to a willowy young woman who's completing a master's degree on medicinal plants, who has a wide, generous smile and drives a 1975 Land Rover with a Confederate flag bumper sticker? Maria hardly spoke English, and she had a boyfriend she would marry by the time I had been back in Canada six months. She wasn't in love with me, but for an evening she humoured me and I let my guard down. I was a long way from home in a place where I didn't know the rules.

We sat at a beaten wooden table in a tavern outside town, drinking *aguardiente* with Maria's friends. I struggled to take in as much as I could of the extraordinary performance in front of us: a three-piece band of the *llano*, Andean harp, guitar and a substantial brown-as-leather woman who was belting out song after song. Love, betrayal, cows, it didn't matter. The singer never stopped moving and the harpist was virtuosic, stamping and ripping off arpeggios. The beat dislocated into a long syncopation and I fell into a state of suspended time where bar lines become obliterated and I could tap my own foot to any number of alternate time schemes and they all made sense. These guys, whoever they were, were brilliant, and I would never have found this place on my own. Maria Lourdes was up and dancing before me.

I rarely dance, except in my imagination, but *aguardiente* is a dangerous drink. A week earlier, in Cartagena, I had watched a soft and spoiled-looking young Colombian consume a string of them at the poolside while lobbing ice cubes at his sulking new wife. She was lovely; only the *aguardiente* would explain such treatment. Now I had had enough of the tiny glasses to cross the same line. I stood unsteadily in front of Maria and concentrated fiercely on her feet, which were moving rapidly while the rest of her body seemed to float in graceful repose. Her legs were like the needles of a sewing machine dialled to produce a complicated parallel stitch. I limped along with her like a rusty Singer, my own movements clashing horribly with the controlled fierceness of the music pouring over us. I forgot about kidnappings and car crashes and let myself go. Off in the distance, I heard someone shouting in English, "Earth to Brian! Earth to Brian!" Another Canadian biologist I was travelling with had clearly had enough of my foolishness. I waved and danced on.

The trip to Villavicencio was about fish. Colombia has the Magdalena River and the Magdalena has fish, somewhat fewer than it used to thanks to agricultural inroads and the peculiar-to-Colombia damage done by the herbicides poured onto remote coca fields by the Colombian military. This is what I knew in 1995: native fish were disappearing, and some of these species had potential value for fish farming. Even back then, people were making the connection between the decline of wild fish stocks and the slightly sleazy role of aquaculture as their "saviour." Fish farmers, even the most clueless, eventually realized that you can't rest your entire operation on just a few original parents, so they began reluctantly to search for a way to broaden their genetic base. Back in 1995, this was the door through which I happily walked, swinging my portable gene banking kit. My job was to help set up a gene bank for *cachama,* a big, ugly river fish with a bulldog underbite and a taste for, of all things, fruit — the same fish I would meet years later in the São Francisco River, where Brazilians call it *pacú.*

There was an aquaculture station in Villavicencio, but the place didn't feel very watery. It seemed about as far from salmon and clear mountain streams as you could get, trapped in stifling heat beneath the foothills. But we all have to start somewhere, and gene banking seemed to me a pretty good place. I'd spent two years developing a way of doing it in the field, out of a plastic suitcase, an applied scientist if ever there was one. I certainly hadn't convinced everyone in Canada to take it up, but there was some genuine interest in Colombia, so that's where I would go. The link to aquaculture, which I could take or leave, went like this: so what if the genetic material you collect could end up in a fish farm? It's also priceless as a living snapshot of wild biodiversity. Surely it's better to put this stuff in a gene bank before it vanishes, and use it later to rebuild the wild stocks? I don't think that way anymore, but in 1995 the reasoning seemed win–win. It had already gotten me a dancing lesson and a spectacular hangover.

Maria Lourdes saw me graciously off at the hotel the next morning, and of course I never saw her again. (Six months later I received an engraved invitation to her wedding.) Adriana, who took her place, was stocky and evidently well-rested and her Land Rover — did all the women in Colombia drive Land Rovers? — lurched into traffic so violently I would have left my breakfast behind if I'd eaten any. She yanked us immediately up a side street, pulled another hard turn onto a red dirt road, accelerated to pothole-leaping speed and drove straight into a rusty iron gate. The gate shook but stood its ground. Adriana reversed savagely and we barrelled back through our own dust.

"No hay via!" she yelled, short hair whipping. No road! Adriana was nothing like Maria Lourdes. I could not imagine her dancing. For the next few days, we travelled between field stations and shadeless, stifling aquaculture ponds bulldozed out of the earth, visiting places so beaten down by budget cuts that the one operable car battery had to be lugged from the Land Rover to the tractor if you wanted to drag some nets out to the fish ponds. Adriana said *"No hay via"* a lot.

Now she was taking us in search of a piece of equipment I had stupidly forgotten to pack, a thing called a canister. This oddly shaped metal sleeve fits the inner contours of a cryogenic storage tank and keeps all the carefully numbered frozen sperm samples from sailing off on a sea of liquid nitrogen. The tank itself was fine; we'd just filled it with liquid nitrogen the day before in Villavicencio. (A cattle town is the one place you're sure to find liquid nitrogen, because most of the cows are inseminated with frozen sperm.) But a tank was no good without a canister to lower into it because you'd be chasing semen samples around the bottom of a narrow-necked vessel in the dark, at -196°C. So, we needed a canister. No canister, no gene bank; no gene bank and I would be spending three weeks at 38°C with nothing to do but twiddle my thumbs and sweat. But this was Colombia, and Adriana had a friend.

"No hay via!" She wrenched us out of another rutted cul-de-sac and slapped the wheel. "Down there," she yelled, waving over a patchwork vista of rolling hills intersected every kilometre or so by fences that meandered out over the horizon. Wherever "there" was, it had to be big. Adriana spoke English of a sort.

"I talk," she said, thumping her chest. "You ..." She pantomimed Canadian politeness, wriggling in her seat and arranging an imaginary tie, possibly a cravat. Thirty minutes later we weren't passing small farms any more, the fences on either side of the road having reduced to a single uninterrupted ribbon that extended much farther than I could see. Whoever we were visiting clearly owned a sizable chunk of the *llano.* Finally we rolled up to his front door, a wrought-iron gate of Hollywood proportions with the words *"Finca Buena Vista"* on a handsome sign ten metres above the ground. A scrawny burro nosed around the scrub. Adriana cut the engine.

"You know why she is called Maria?" Adriana reached past me, opened my door and squeezed out of the Land Rover. "This *burro?*" She rolled her tongue around the word and scratched the animal behind one ear. "Is the name from the men, right, *chica?*" The burro flicked the skin on its back and dislodged some flies. "All day long, work in the field, at night, the men are so lonely." Adriana stuck her tongue out at the animal, which looked as embarrassed as I was. "Okay, just a moment." Adriana walked over to an intercom set into one of the stone arches and leaned close to speak into it. Her body language began to change. She seemed a good deal more deferential. She motioned us back into the Land Rover.

"Okay," she said. "He has. We go." The gate swung noiselessly open and Maria the fallen burro stepped incuriously aside. The gate closed behind us and I looked up and saw an antelope.

It didn't faze me at all. I had already found Colombia to be a crazier place than most; suspension of disbelief was becoming a habit. In my hotel in Bogotá, I had heard what sounded like gunfire. Asking around the next morning, I determined that it was. Nobody appeared concerned. So: gunfire was normal. Suspension of disbelief. Braking for an antelope in the middle of the *llano* triggered the same "anything can happen" response. An antelope . . . so? The animal bounded away and we continued. Something with long twisted horns and melting eyes peeked at us from behind a tree. Then, farther off and behind a fence, I saw the glint of water. A hippopotamus stood in it. There were also flamingos. We were driving through a private zoo.

"*Narcotráficos,*" whispered Adriana. Two brothers; she told me their names and I decided to forget them. Both were associates of Pablo Escobar, that great benefactor, a hero to his people. Escobar had just been arrested, allowed to escape, recaptured and eliminated, a procedure that saved the country millions in legal costs but had no effect on the flow of drugs out of Colombia. Neither brother would be able to greet us personally on account of being in prison. However, their private veterinarian was available.

We passed a netted aviary that would have fit nicely into a football stadium. Farther on, the helipad and then the house and finally, behind the swimming pool, the reason this particular vet might have a canister we could borrow: the family bullring. The brothers, it appeared, bred their own. For which, of course, they would need access to the finest frozen semen on the global market, for which, in turn, their vet needed canisters — including, we hoped, a spare one. Weeds poked up here and there in the dusty central pit but I could imagine the family and their friends filling the little grandstand, lifting their drinks while the children scampered between the rows of seats and the bulls pawed at the dirt. By this time I hardly needed to suspend disbelief at all. It was doing fine on its own, hovering up above the trees, well out of reach.

First, we sat down outside, in the shade of the house, beside the pool. The surface of the water was littered with leaves. The vet was cautious-looking, unsmiling. He offered coffee; we indulged in some territory-marking while two men watched from the corners of the house. The vet had Ivy League training and spoke good English and he carefully considered the names I was dropping. Finally he nodded imperceptibly at one of them, demonstrating once again the absurd connectivity of everyone on the planet: five minutes with an indentured vet on a jailed Colombian drug lord's farm and you've established you both know the same obscure academic in Winnipeg.

But I was cleared. Adriana made a relieved little noise and I gazed at the leaf-clogged pool while one of the men was dispatched to fetch the canister. The vet lapsed into silence and we sipped our coffee. Somewhere on the grounds there were microscopes and liquid nitrogen and special stalls for confining the animals for treatment. Maybe the *narcotráficos* had even invested in a mechanical cow for collecting semen from their prized bulls. I'd seen one in a French catalogue; the thing had wheels and you sat inside the cow's butt, holding a bucket. I couldn't imagine a Colombian male agreeing to do this.

Finally the assistant returned and surrendered the canister solemnly; his look hinted of the consequences of not returning it. Hands were shaken, business cards exchanged, and we made ourselves scarce. I peeked in a window on the way back to the car; all the furniture was draped, like so many coffins. A man trailed us back to the car and we trundled back down the long drive. The hippo hadn't moved but the ibex or whatever they were had vanished.

The gates opened. Maria was still cropping at the weeds outside. Maybe she was building up reserves for the night ahead. Adriana became her voluble self again, and I tried to share her relief at coming away with the canister but I already knew what would happen next: some waiting around a pond, a lot of splashing, and the Colombians would have the first samples in their *cachama* gene bank. I should have been happy, but all I could think about was getting to a flat stretch of highway so I could pull out my notebook and write down what had just happened. Who cared about gene banks when I had just sat by a *narcotráfico's* pool?

Fish, science, aquaculture, Colombia: a potent mix, even without the *narcos*. I didn't realize it at the time, but all those hatcheries — the sweaty South American ones, with their inscrutable catfish, the lonely Canadian ones where the salmon blood ran under the dripping cedars — were the most volatile ingredient. Hatcheries are used not just in fish farming but also to restock wild species, and the time I spent hanging around them had an unexpected result: it allowed me to see how closely fisheries and aquaculture were becoming bundled together. The links were in shared technologies, long-term planning, and what people were starting to call "food security," and they were links a lot of people didn't want to acknowledge. Clearest of all was the amount of control we now had over fish — farmed or wild.

Where would all this control take us? Does the ability to harness nature and micro-manage it provide lessons for what was happening to the São Francisco River? To find out, I had to visit a country as far from South America — geographically and culturally — as you could probably get. I went to Japan, where planning is everything. And I met Miss Tojo.

CHAPTER 9

MISS TOJO'S TUNA

Managing nature in Japan

Gone in sixty seconds: inspecting tuna for auction at Tsukiji Market, Tokyo.

LIKE MOST PEOPLE, I GO TO BANFF FOR THE MOUNTAINS, not for the sushi. But after a day of scrambling up scree slopes and punishing your knees on the descent through the treeline, there is something comforting about pulling up a stool around a miniature railroad towing colour-coded plates of raw fish past your nose. It's called *kaiten* ("around and around") sushi. Banff has only one such shop, but the food is good, the place is relaxed and usually full; that's where I like to go. The chef stands encircled by the track and his customers, cutting, scooping, rolling, and placing his creations on model boxcars that say "Canadian National" and "Great Northern." Occasionally he takes a special order or yells for reinforcements from the kitchen and an assistant ducks under the tracks to pop up with a plastic tub of sea urchin roe or a fresh slab of *hamachi*. The Japanese are good at yelling, in the approved circumstances.

Hamachi means yellowtail, a muscular amberjack that looks like a scaled-down tuna. On my last trip to the mountains I watched the chef unwrap a loin of it. He was young, surely no more than twenty, but his movements already had that balletic blend of the rapid and the deliberate that you see in all good sushi chefs. The piece of fish was pearly, about the size of a rolled-up newspaper, shot through with delicate pink striations. He cut two segments from one end, the pale flesh falling smoothly away from the blade, and draped each over a crooked finger of rice. The rest of the yellowtail went back into the chiller beneath the counter and I followed it with my eyes while the two pieces of sashimi started around the circle. Yellowtail is delicious; by the time the little boxcar reached me they were gone. *Damn*.

Where did this exquisite piece of flesh come from? How was it possible for a hiker in Alberta to pop it confidently into his mouth? That slab in its plastic wrapping, how absurdly far it seemed to me from the place where it was caught. Freezers, yes; jet aircraft, quality control — we all know of these things, although rarely in any useful detail, and we trust them every time we go to the supermarket. What interested me about the yellowtail was that it represented something more than just an efficient and reliable mechanism for taking an animal out of a fishery and delivering it halfway around the world. What it meant to me was the absolute determination to ensure that yellowtail would always be available, not just to tired hikers in Banff but to housewives in Osaka. That yellowtail was from Japan, and it stood for a state of mind, a very interesting one for the future of fisheries and perversely relevant to the São Francisco.

Almost certainly, that yellowtail was farmed. In Japan, the concept of a "fishery" has gone as far as it possibly can from the lone São Francisco fisherman clinging to a rock with his toenails and slinging a net round his head to bring in his daily sustenance. In the old days, Japan was like that too. But yellowtail this fresh and affordable has to come from a chainlink cage, suspended in a quiet bay and machine-gunned twice a day with pellets from a computer-controlled feed barge. Professional traders and chefs may regard farmed seafood as inferior to wild, but the average Japanese shopper grudgingly accepts most farmed fish as a necessary evil.

Sometimes it's even better not to know, because it may take the fun out of eating. That's the case with *fugu*, the symbol of the Japanese obsession over those qualities of food that transcend nutrition. *Fugu* can kill you, everybody knows that: the slightest wandering of the chef's razor-sharp blade can graze one of its organs, releasing the tetrodotoxin that will stop your heart. When I saw a tank of *fugu* in a friend's laboratory in Japan I asked him about this and he hung his head. "Um, farmed," he told me sheepishly. No wild bacteria in their diet, so no toxin. The wonderful slogan "Farmed and Dangerous" works for environmental groups campaigning against salmon farming, but with *fugu* it's just the opposite.

Fish cages, Amami Island, Japan.

Fisheries agencies love to quote statistics: so many tonnes taken, so many left, this sector depleted, this one ripe with promise. Markets appear and disappear, and new technologies promise to bail us out of whatever biologically untenable situation we create. But even the word "fishery" fifty years from now may be an anachronism, because more and more species will be farmed. The future is about razor's-edge management of wild stocks and a slow, rolling tsunami of farm-grown fish, and not only in developed countries. The São Francisco too will see the balance tip; it's already starting to, with fish cages bobbing up in the Três Marias and Sobradinho reservoirs, even in slack stretches of the river itself. If you want to know what's in store for fisheries and rivers in Brazil and a lot of other countries, there's a place where you can already see the future: Japan.

Once having made their minds up, the Japanese never do things by halves. It's "minds" because Japan is a collective society that operates by consensus. For a Westerner, Japanese consensus can slow decision making to an infuriating crawl, but once a decision is made, watch out. If individual Japanese often seem paralyzed by the need to operate in harmony with everyone else, the Japanese determination in carrying out a group decision is legendary. Remember: after the Second World War, the Japanese rebuilt their major cities in a decade, and they had had some practice already, having created virtually a new Tokyo in seven years after the earthquake and fire of 1923. Walk down a street and look at the space between the office buildings: in many cases it's less than three feet. How do you insert a forty-storey structure into such a confined space, let alone demolish the previous one once it's deemed too old and out of date? The Japanese do it all the time. Only eight months after the Kobe earthquake of 1996, the city had already been stitched back together, the only signs of devastation being the temporary housing and the foot-wide cracks in the streets that ran like jagged scars where the concrete had split. So never underestimate the Japanese when it comes to fisheries, because they decided thirty years ago to do momentous things there, too. And once we understand that attitude we can recognize it when it pops up in other countries — including Brazil.

FISH IN MY FAMILY

My wife, Hatsumi, is Japanese. Her father spent his adult life running the family business in Tsukiji, still the greatest fish market in the world. Unfortunately, neither of these facts makes me an expert on Japan. Quite the opposite, as I am reminded daily; in my relationship with Japan and the Japanese I am one of those people who knows just enough about something to realize he knows practically nothing. For me to speak the language is to skate on the thinnest of ice; I usually crash and sink only a few strides from shore. But I do know the culture in the way that can only come from living with one of its human products, for a culture need not remain at home to express itself. At times it even seems the opposite: the farther from home, the stronger the pull of the customs one grew up with. Another way to say this is that learning to step over a collection of shoes in the hallway may be a better cultural immersion than a hundred hours with *501 Japanese Verbs*. It's definitely better for your coordination.

My father-in-law has been in the fish business all his life. I don't think he's a "typical" Japanese; for one thing he's taller than I am. He's not a typical fish merchant either; he has a degree in political economy from Waseda University, the second-most prestigious in Japan (number one is *Tōdai,* Tokyo University). But when the family wholesale business in Tsukiji Market required someone to take it over, my *otōsan* ("Father") was the first son, so he did it. He had no choice. Maybe that's typical enough.

Otōsan has a large, commanding head with movie-star grey hair flowing high off his forehead. His lips are generous, and sometimes he pulls at the lower one and makes his "thinking" sound, a guttural, drawn-out "Ohhh" accompanied by a frown I would not like to see if I were one of his employees. When he is perplexed he sometimes scratches the top of his head. In Tokyo, the place my wife refers to as her "home town" and to me is an incomprehensible concrete horror I can't escape from soon enough, *otōsan* stands out. His clothes seem to have been chosen from the wardrobes of several different personalities: jeans with a cream-coloured silk vest; a suit with sandals; rose-tinted aviator shades and a bolo tie.

He and my *okāsan* ("Mother") live in the house they built fifty years ago in Kugayama district, hemmed in to near-immobility by all the things they have never been able to throw out. Life revolves around the tiny kitchen where *okāsan* prepares food on a handkerchief-sized stainless-steel counter next to a two-burner stove while *otōsan* holds court at the kitchen table, surrounded by wobbling stacks of books, maps, old photos, fishing lures, the small television that flickers with news and sumo matches all day, even a leaning tower of hats and sun visors that one could go through, if one had a mind to, layer by layer, back through decades of Japanese style. Underneath, your knees run into more books, a dimpled foot massager, a wooden box that opens out to become a tiny vanity where *okāsan* kneels every morning to do her makeup. I think there's a chest of drawers under there too, just a small one.

The last time I visited, I awoke at five each morning to a drumming sound that seemed to come from directly beneath my futon.

There were other sounds — the ubiquitous Tokyo crows, sound-trucks promoting local businesses and political candidates — but the drumming was loudest and it seemed to come from the bowels of the earth. My eyes popped open like a doll's and I reached out to touch the wall with my fingertips: an earthquake!

But none of *otōsan*'s exquisite Chinese dolls flew out of their glass cabinets; the faded fish-print of the eleven-kilogram snapper he caught in 1986 didn't pop off the wall and roll itself up. After the third day of this I dragged myself out of bed, cracked open the sliding paper door and peered down the stairwell to see my seventy-three-year-old father-in-law in his royal blue dressing gown, upended on his shoulders and drumming on the floor behind his head with his toes. After that he sat, shirtless, at the kitchen table, watching the news and checking his blood pressure. He wrote the numbers down in a ledger that contained his cardiac history for the past thirty years.

I was still awake an hour later when the door of the Toyota Crown slammed and *otōsan* eased the big car out of its tiny garage and shot down the narrow alley toward the Sumida River and the market. I turned over and went back to sleep imagining him adjusting his aviator shades, gunning into on-ramps, rummaging blindly in a jumble of plastic cards for the one that would open the electronic toll gate at the last possible minute — *ding!* Maybe he would fumble with his left hand for the electric shaver he kept in a zippered pouch in the glove compartment, whipping the Phillips around his chin while the Toyota split the seams in the traffic. He would peer intently at the electronic navigation system whenever the car came to a stop and the screen switched automatically to television, then release the handbrake with a bang when the light turned green. If the Tokyo traffic were one of those silver shoals of anchovies, *otōsan* would be the fish that moved first.

This is a man who knows his fish. The Japanese love of seafood is a cliché done to death many times over, and it's full of misconceptions, such as the one that has the Japanese eating more fish per capita than anyone else. Yes, they eat three times as much as North Americans do, but they're still well behind a lot of small island nations like Greenland or the Faroes. The *real* fish-eaters are in

Oceania, in tiny nations like Palau or Tokelau, where people consume nearly four times as much as the average Japanese. Nevertheless, being a part of *otōsan*'s family has allowed me to see that the Japanese love of seafood, like other kinds of love, needs some qualification. It is not a gentle adoration. It's a devouring love, one-sided, the kind that can never be satisfied.

The seafood that covers *okāsan*'s kitchen table once the books and hats have been pushed aside, whether it was ordered in or prepared by my in-laws themselves, is not simply eaten. It's *devoured*. Chopsticks and fingers reach; crab legs are mangled and sucked; pungent slivers of dried mullet vanish, lips smack. The Japanese — people like my father-in-law and the customers he serves and the policy makers and business leaders that listen intently to what the market tells them — long ago decided that there was no way their country would ever run out of the kind of food I can eat at that kitchen table. They made a collective decision, as determined as the vow to rebuild Tokyo after the Allies fire-bombed the city flat, an event *otōsan* remembers as though it happened yesterday. That decision has implications for every country with aquatic resources.

THE MOTHER OF ALL FISH MARKETS

Food is a serious business in Japan, especially given the isolated nature of the archipelago. And being a connoisseur of seafood is just one of many cultural duties for every Japanese. Seafood is celebrated in Japan through books, TV programs and cooking shows (*Iron Chef* is just the tip of the iceberg), endless magazine articles, even *manga* (comic books) about being the world's greatest chef. Juzo Itami's wonderful film *Tanpopo* pokes fun at the food mystique by creating an endearing heroine obsessed with learning how to make the perfect bowl of *ramen*. In one scene, a man upends a glass bowl containing two live shrimp on his lover's belly button and the two of them gasp and giggle while the crustaceans try frantically to swim their way out. In another, the two perform what has to be the silver screen's most glutinous kiss, passing a raw egg yolk back and forth until it bursts. Only in Japan.

Food symbolism — the way a live eel is pinned through the head

In Tokyo, buying seafood can be an education.

and skinned, the auspicious colour of *tai* or sea bream or the popu-
larity of sardines as a good-luck food in spring (devils don't like
their smell) — is deeply embedded in Japanese cultural history.
Seafood lore is part of Japan. Here is my favourite: there are few
female sushi chefs because women have warm hands and would
therefore ruin the delicate flesh. This one of course is nonsense. My
wife frequently has cold extremities; furthermore, the best sushi
restaurant in Victoria is an all-woman affair. Silliness aside, Japanese
people care far more deeply about their seafood than do most cultures, and this goes beyond appearance and season to method of
preparation, aesthetics and health benefits.

That sensibility can make it next to impossible for exporters to
crack the Japanese market, because other cultures have trouble with
the Japanese requirement for appearance and arrangement. This is,
after all, a country where you can go to the food section of a department store and pay $300 for a melon that tastes like any other but
whose shape, colour and texture are suitable for presentation as a gift

of the highest significance. The North American salmon exporter who tries to send a shipment of scarred or blemished (but still perfectly fresh) fish to Tokyo finds out about such sensibilities in short order. A few years ago, when the demand for wild herring roe on kelp was fuelling a cottage industry in British Columbia's coastal communities, the biggest hurdle for the locals was the Japanese buyers.

"These guys are just crazy, so picky," one Indian woman told me. "Everything has to go through one buyer in Vancouver, and it's gotta look right." No wonder; *kazunoku konbu* is a traditional gift around Christmas time; your recipient is going to know *everything* about it, and it had better be wrapped just so. It was the same thing with *iwashi* (sardines): BC had plenty of them but few fishermen were prepared to jump through all the hoops necessary to please a Japanese buyer.

If there is one acknowledged symbol of that one-sided love affair, it's the Tsukiji Market. Tsukiji is central to Japanese culture. The media present the market as the "village" whose denizens are custodians of the cultural role of seafood — somewhat quaint, barely tolerant of outsiders. I don't know about quaint, but the barely tolerant part is accurate. A clumsy *gaijin* is even less welcome than a little knot of Japanese housewives who've decided to brave the market for something really special. The Japanese are a very polite people, but you can generate lots of dirty looks if you step the wrong way in Tsukiji.

If you enter the outer market from the southeast corner, that is, from the direction of the inner market, you first walk past Namiyoke Jinja, a small Shinto shrine. Its signs and banners bear the names of Tsukiji firms, and it's best known as the place where seafood is thanked for yielding up its collective life to feed the Japanese. Atonement is important: merchants of Tsukiji may spend their days surrounded by dead and dying seafood and a few of them are gruesomely involved in butchery of living fish, like the eels that are pinned and skinned alive. But their religions provide for ritual pathways to forgiveness, and the shrine at Namiyoke is the entrance to that path. The shrine is less famous as the place where I proposed to my wife (more accurately, I popped the question in the middle of a crosswalk after viewing the shrine, but that's somehow less romantic).

Tsukiji is the biggest fish market in the world and its dealers buy

Inside the Tsukiji market, Tokyo.

and sell more than four hundred and fifty species of seafood. Tsukiji likes to present itself as *Tokyo no daidokoro* (Tokyo's pantry) but if this is a pantry it's a huge one, handling almost ten times the tonnage of North America's largest, the Fulton Fish Market in New York. It's also a very international pantry, because its goods come from every ocean. Amazingly, most of the trade that flows through Tsukiji is handled by tiny family businesses like my father-in- law's. In Tsukiji, family businesses connect with multinationals. Trade buyers like *otōsan* are bridges between the two.

The impact of the place is visceral. I went there the first time with my friend Minoru, then a scientist working for the largest fishery corporation in Japan. This was years before I became part of a Japanese family, but the memory of that early-morning visit persists: glistening concrete floors and a warren of wooden stalls lit by fluorescent tubes and low-swinging lamps; whole deep-frozen tuna laid out in smoking rows and auctioned at lightning speed; eels undulating in their own blood. Skinny men wielding knives like samurai swords deconstructed bluefin tuna into gleaming purple segments the size of a man's leg while, in a corner, two exhausted-looking men in rubber boots took a guilty tea-break beside a fire built in an empty pail. I remember the noise of the band saws, the incomprehensible array of species and the incongruous sweetness of the air.

I also remember Minoru yanking me out of the path of one of the motorized delivery carts that race up and down the narrow aisles, driven by stone-faced men in hard hats. But what really stuck with me was that it was possible to bring so many tonnes of fresh marine life into such a packed space with such efficiency that, if you closed your eyes, you might just as easily have been in a warehouse for some other commodity entirely, shoes, say, or televisions. It was that clean.

But Tsukiji is more than just a place to sell and buy fish. It's also a cultural institution with a long memory, back through feudal Japan under the Tokugawa regime, the emergence of an aggressive mercantile class, restoration of the emperor, earthquakes, fires, wars and rebuilding. In a city where outside cultural influences are all over the map, where you can be driving past a row of modest shops and suddenly see a Maserati on a turntable in a glass showroom next to a doggy hotel, enduring icons like Tsukiji take on great significance.

In the early seventeenth century, Tsukiji was just a marshy landfill in rapidly expanding Edo (now Tokyo), the seat of the Tokugawa dynasty. The shogunate had its own fish market in Nihonbashi, on one of Edo's many canals, on the present site of the Mitsukoshi department store in the centre of banking Tokyo. In feudal Japan, the Nihonbashi merchants depended on the shogun, and they passed that dependence back down to their suppliers, the fishing communities; the regime's right to the best seafood thus extended all the way to the villages where it was caught. Despite the merchant's monopoly there was no escaping the obligation to supply the court, and at bargain prices.

A true mercantile economy didn't have a chance to emerge until the new Meiji government abolished the shogunate's control in the 1860s. Tsukiji district used to be the Foreign Settlement, an area created by the Meiji government to keep the inevitable outsiders out of sight. The area was made the site of the new fish market when

After the auction: otōsan leaving the Tsukiji market. Note the boots.

the Nihonbashi market was destroyed, along with most of the rest of Tokyo, in the great Kanto earthquake of 1923. But the traditions of the old mercantile *shitamachi* district — work hard, play hard, small family firms — survived the move to Tsukiji. The Foreign Settlement itself, an enclave of diplomats, teachers and missionaries, lasted only a generation.

Nowadays, most of Tsukiji's seafood arrives by refrigerated trucks that begin rumbling in around ten at night. Restaurants open, and unloading goes on until the auction pits are ready to open around five the next morning. Over the next two hours, twenty-three hundred tonnes of seafood change hands and are artfully displayed in the curving rows of 1,677 stalls, ready for the buyers at seven. It's all over by ten in the morning.

There are really two markets in Tsukiji — inner and outer. The whole complex covers a city block. The inner market is decaying Bauhaus, a small city of parallel stalls bordered by the twenty or so auction pits, of which the one for tuna is the most famous. The inner market is where seafood is auctioned and resold; the outer market, where my father-in-law has his shop, is a warren of whole-sale and retail shops, restaurants and small office and apartment buildings. It's as far as most ordinary Japanese penetrate into Tsukiji. You can easily imagine the black market that bubbled along here in the years after the Second World War, when Japan struggled to rebuild its fishing fleet to take advantage of one of the war's few silver linings, the abundance of fish in the Western Pacific. *otōsan's* wholesale business is a small space with upstairs shelves stacked with packaged foods and, downstairs, a cold-storage locker filled with processed fish products like fish cakes and *kamaboko* (fish paté), all purchased in the inner market by virtue of the all-important licence he inherited from his father. There's barely enough room to swing the mini-forklift. *Okāsan* sits in the back with an adding machine and a stack of ledgers.

FARMING THE SEA — INCLUDING TUNA

Tsukiji is a marvel, yes, and has remained so for me on subsequent visits with *otōsan* — whose stature as an insider, I don't mind saying, brings added satisfaction (and safety). One would have to be an idiot to miss the message of Tsukiji, and later we'll see how it applies to grandiose plans halfway around the world, in Brazil. But it's not the best evidence of the choice the Japanese have made. To see that, you have to leave Tokyo and go to where the fish are. Which is to say that, more and more, you have to leave Japan. From a nation that was, thirty years ago, self-sufficient in seafood but for a few imported oddities, Japan has become a net importer and its seafood industry has changed dramatically. It's no longer a home-based network of vessels, villages and smoothly functioning cooperatives that kept the fresh fish flowing through Tsukiji. This is now a nation that relies on a many-tentacled industry in which not only do Japanese boats operate far offshore, but deals are also made with other nations for access to species that have been fished out closer to home.

Often, these deals are with developing countries and reflect the aggressive economics of any resource sector. Parallels can't be drawn too closely; the Japanese would never dream of going to war over seafood the way some nations have done over oil, but then again they don't have to. They already have agreements with seafood-rich countries, skilled and determined representation in all the international conventions related to marine resources, and direct investment or massive foreign aid tied to economic deals. When the Director of Fisheries for FAO is an ex-representative of the Japanese fishing industry, you're covering the bases.

Japan is a maritime nation: four major islands and three thousand smaller ones, an archipelago in the middle of the highly productive oceanic mixing bowl produced by the Kuroshio and Okhotsk currents. The ports that dot this immense coastline are mostly small, and serve local fisheries; the few dozen big ones serve the distant-water fleets managed by the vertically integrated companies like the one where my friend Minoru worked. When the Tsukiji market started up in 1935, its docks on the Sumida River were where the ships and the trains arrived, and both were full of fish from Japan's own waters. Although the seafood arrives mainly by truck now, the real change is in how many of those trucks pick up their cargo at Narita International Airport.

Nowadays, dependence on imported seafood is huge, but even defining "imported" is tricky: is the tuna caught by a Japanese vessel in the Indian Ocean or off Chile imported or "Japanese?" *Unagi* (eel) are 100 per cent imported, mostly from China; 80 per cent of the *uni* (sea urchin) are from somewhere outside Japan. I can remember a boom in *uni* from my own home town, Victoria, the divers cleaning out the subtidal rocks and piling the crates of fresh *uni* right beside the road. It was a brief boom; Japan moved on.

But the deals, the subsidies, the lobbying are not unique. All rich nations have them. What's more interesting about the Japanese strategy isn't the way people have managed to keep getting hold of seafood using nets and hooks (even if those nets and hooks belong to someone else), but how they have rethought the whole business of fish and come up with a radically different solution: aquaculture. The Japanese made a commitment to aquaculture long before the average person, in any country, was aware that fish farming even

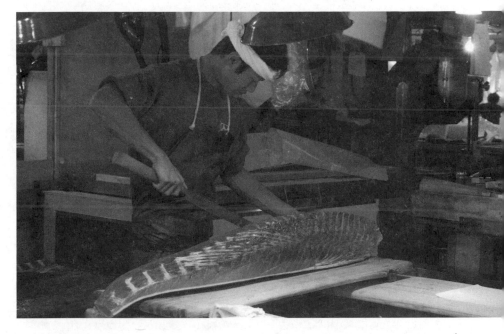

Disassembling a bluefin: but where did it come from?

existed, long before anyone cared about whether the fish they were eating was farmed or wild.

Japan was certainly not the first to go this way. For sheer tonnage of farmed fish, China has led the world for as long as statistics have been kept, and China still produces around 70 per cent of the total volume of farmed fish and shellfish and aquatic plants. Aquaculture is so dominant in China, in fact, that more than twice as much fish is farmed than is caught the old-fashioned way. The Japanese wrinkle was that the aquaculture option was adopted nation-wide, not for export but to serve the needs of its own people, and the decision affected every layer of society. Like the rebuilding of Tokyo.

If you were a fisheries scientist, you saw it coming. Back in the 1960s and 1970s, it was hard not to notice that an awful lot of the scientific reports about fish were in Japanese. It was frustrating: you'd run across the title of an article and think, *Yes!* The very thing I've been looking for! And then the rest of the citation: a long list of Japanese names, the name of the journal (again, Japanese), and the final nail, the dreaded "in Japanese" in brackets. Foiled again. Japanese society funded the researchers, the researchers toiled, and the national game plan for seafood unfolded. Is still unfolding, I should say: the other day, I ran across a Japanese fisheries paper titled, "An Automatic Chasing Robot Boat for Salmon Migration Research." I think the idea here is that the mystery of salmon migration could finally be solved, if only we could catch them at it.

It wasn't just the universities doing the research, because the big fishing companies stood to gain spectacularly. My friend Minoru, the one who took me first to Tsukiji, is a good example of the plan. Minoru is fire hydrant–shaped guy with thinning hair and a metallic grin that lights up his face. Laugh lines radiate from his eyes. Twenty years ago, his letter appeared out of the blue, asking if he could visit my laboratory and travel around looking at other labs in British Columbia. Sure, I wrote back; I was flattered. Minoru worked for a big Japanese fishery conglomerate. This was my big break: lucrative research contracts and all-expense-paid trips to Japan! Minoru told me how to recognize him at the airport: short, a red bag, a moustache. At his hotel the next morning he was the same height but the moustache was gone.

"I grow for meeting only," he said. Minoru, or maybe it was his company, left nothing to chance.

I was never really sure what Minoru did back home, although I still think of him every time I drive past Nanaimo, a city on Vancouver Island that's the jumping-off point for a number of fishery operations and research efforts. "Seven potatoes," Minoru intoned as we passed the sign on the highway. Of course: *nana* means seven, *imo* is potato. Seven potatoes, I still call it that. After that first visit we kept in touch and I even visited him once at his laboratory in Tsukuba or "Science City" outside Tokyo (he was the one with the farmed *fugu*), but we never made any concrete plans. But about ten years later he called me from Japan and asked for my help again, and this time he was specific: he wanted to freeze tuna sperm, and he wanted to do it at sea. Of course he had to do it at sea, because there weren't any tuna in captivity, not the big monsters who were sexually mature. If I helped him with some suggestions, he would even put my name on the paper he wrote about it.

This was interesting. Tuna, Minoru bobbing around on a boat with his own little laboratory — I figured out what was going on. You freeze sperm when you want to use it later (sometimes decades later) for controlled breeding. It had been done for farm animals for decades. I had published enough research to get my name known in the field; that's what I was doing on that hare-brained trip to Colombia. When Minoru told me his plans, I sat up and listened. Ten years ago he had visited me, checked me out. Now his company was going after the prize: farmed tuna. And this was a company whose core business was catching them at sea. Planning plus determination, all very Japanese.

"I'll put my thinking cap on," I said.

"Hunh?" Minoru said.

The tough part wouldn't be freezing the sperm; it would be obtaining the sample. I'd removed samples of sperm ranging from microlitres (from a two-centimetre-long zebra danio) to a sleazy-looking half cup from a one-metre black cod, but I knew precisely nothing about the crucial reproductive anatomy of a tuna, and a live one at that: you have to know the size of the testes, the assorted tubing and openings, the chance of getting a faceful of urine by mistake. What would happen with a tuna was anybody's guess, and I decided to leave that part to Minoru and his team. They weren't inviting me on board anyway. I would have to imagine them clustered in an anxious knot on the rolling afterdeck, beakers in hand, white-coated and wearing the regulation bright blue Japanese fisherman's rubber boots. In the centre of the ragged circle of technicians I could see the majestic tuna on its side, eyeing them balefully. The great beast would flex and the scientists would draw together and confer.

I suggested some formulas for protecting the sperm for its headlong journey from the tossing deck to liquid nitrogen, a drop of over two hundred degrees. Amami Island, where the experiment would take place, was tropical, so those technicians would be sweating under their white smocks.

"Glucose," I told Minoru. "And dimethyl sulfoxide, add some of that. When it's time, mix in a little egg yolk, the sperm cells seem to like that too." On the other end of the line, Minoru went *unh, unh,* writing it all down in kanji, and I looked out the window at the bare trees and the rain and wished that, wacky as his request was, I could go with him. But really, what chance did it have of working, my dreamed-up recipe? Better I should stay at home, the embarrassment would be less.

And then, a few months later, I got the draft of the scientific report Minoru was writing and realized that my mixture *had* worked, that he'd done it. His company had added a crucial link to the chain of techniques needed to close the reproductive circle and domesticate the icon of Japanese seafood. There were congratulations and thanks, and then . . . the company dropped the ball. I got another letter from Minoru. Despite his success, they were closing his lab. Minoru was moved sideways, to Ice Creams, where he would be expected to work long hours perfecting new flavours based, I imagined, on novel ingredients from the sea. Who knew, I suggested, maybe he could even add a dash of bluefin sperm, but Minoru wasn't in a laughing mood. In a very un-Japanese move that I've always admired him for, he quit the company and went back to school to get his Ph.D. Our method for the perfect sperm cocktail stayed on the shelf. Nobody flew me to Japan.

Their reason, of course, was economic. Minoru's company had been following reports about tuna-fattening, in which young fish are netted, gently towed to great floating pens, and pampered for a year or more with shovels-full of fresh sardines. Here, the economics were excellent and the technology less risky than formulas obtained from far-away *gaijin*. Tuna farming, or rather fattening, is a fact of life now, in Mexico, Australia, the Mediterranean, Canada, even Ireland. The farms are steady suppliers for the Japanese, who use mathematical models to constantly gauge the merits — cost, delivery time, quality — of each country. The key thing is control: fish are there when you need them, supply is assured. Minoru's company decided to put tuna aquaculture on hold and invest in tuna fattening.

The fattening operations in Mexico use the cool waters along the northern coast of Baja California. It's convenient: most of the product gets trucked quickly north to Los Angeles, where it's put on the plane for Tokyo. Many of the operations have Japanese investors

Fattening caged bluefin tuna, Amami Island, Japan.

and are marvels of efficiency. They're not the full-on, egg-to-adult farming that Minoru's preliminary research was aimed at (don't worry, that's still coming), but you can see why companies like his put their money into herding and feedlots. There's no big investment in risky technologies and no change whatsoever in the number of tuna that get removed from the sea, even if you have to set up a separate fishery for the sardines to feed them. It works. Boy, does it work.

The young bluefin are caught in seine nets when they visit the California coast in summer. They're towed slowly to the fattening sites near shore, a journey that takes several days. I wonder, are they as stupid as their tiny brains would suggest, just streamlined machines for eating and reproducing, or do they know something's not right, crowded together like prisoners in a paddy wagon driven through an everyday street. On the highways in Alberta, when you pass a semi-trailer full of cows headed to slaughter you can smell the shit and urine and I swear you can smell the fear, even in the fleeting exchange of air between a truck going one way and your own car going the other. Do other fish smell the tuna's distress and arrow away in alarm? But the tuna complete their final journey, and unlike the cows they get an eight-month reprieve at the end of it, circling like pashas in forty-metre fattening cages, hand-fed with fresh sardines caught by the company's contractors.

It all ends badly, of course. Divers in black neoprene enter the pens like saboteurs, unfurling a curtain of netting to isolate the first unlucky fish. Hemmed in, the fish are approachable and the divers manhandle them to the team on the killing boat where each fish is swarmed, grabbed, pinioned. This isn't the comical scientific team from my imagination; the job isn't to collect sperm for some flaky experiment but to kill, bleed and chill a two-hundred-kilogram fish — fast. One man straddles the head, finds the unprotected spot between the eyes, inserts a metal spike and leans into it, striking for the brain. It's a small target but easily destroyed once you know how to find it. When the final shudders cease, a stainless steel wire is run into the hole, like the "snake" you feed into a clogged drain. The wire finds the opening to the spinal canal and reams through the chain of vertebral discs, obliterating the spinal cord and short-circuiting all the post-mortem biochemical reactions that produce a lower-quality fish.

Bleeding is next, with a cut to the artery supplying the gills, then these too are swiftly sliced out, a handful of cherry-red epaulets torn from a disgraced officer. The fish is on a table now with a dozen others, slithering along, leaned over by another fiendish group in rubber aprons, face masks, hairnets. Bellies are ripped, guts twisted out and flung aside, and the tuna slides off the end of the table into a chilling bath the colour of burgundy. Time elapsed: less than a minute. At the peak of the season the killing boats go through hundreds of fish a day.

The rest of the cadaver's journey is anticlimactic: by boat to the packing plant, then measuring, weighing, wrapping in plastic and gel packs and the short ride north by truck to the All Nippon 747 waiting at Los Angeles International. Within forty-eight hours of the entry of the black-suited divers into the fattening pen, the bluefin are lying in rows on the spotless concrete floor of the auction hall in Tsukiji market, numbered and tagged and with an ugly notch in the tail where the prospective bidder can shine his flashlight, checking out the flesh. My father-in-law, if he got up extra early and skipped the foot-drumming, might bid on one.

But this whole ghoulish system is just a stopgap. It's just processing, really, fundamentally no different from fishing. The fish still have to be caught the old-fashioned way, and if quotas go down in one area or lax local regulations get tightened, they have to come from somewhere else. This is the eventuality all those Japanese-language research papers written thirty years ago were anticipating, and the resulting strategy is fish farming. Cradle to grave: eggs, sperm, babies, growout, harvest. Sometimes you can persuade the fish to breed naturally in cages, perhaps giving the breeders an extra little push by injecting them with neuropeptide hormones that kick-start the gonads into producing ripe eggs and sperm. This step has already been achieved with bluefin. Larval survival is often a problem, but researchers are inventive there too, figuring out ways to grow enormous vats of the microscopic plankton that baby fish need to eat.

The history of aquaculture development in Japan, where most of the farmed species are high-value marine ones, has been the dogged development of fixes for all the problems of spawning and artificial insemination and larval rearing. One by one, the "impossible" species have been domesticated. Oysters, scallops, sea bream, certainly the yellowtail in the *kaiten-sushi* shop in Banff. Bluefin tuna too; it's already being done on a pilot scale in several labs. If you're driving along the seacoast almost anywhere in Japan and you come across a protected bay, somebody will be farming something. The walkways around fish cages undulate uneasily on the surface like a carpet of enormous hula hoops, and fluorescent orange buoys mark longlines hung with trays of oysters and scallops. From the air, pearl farms look like the lights of an airport runway. Plants too. The Japanese cultivate entire underwater forests of kelp and a dozen other seaweeds using methods published — you guessed it — in all those obscure journals. Except that they aren't so obscure in Japan, and the Japanese are still eating their seafood. Planning, again. If you know how to cultivate enough kinds of seafood you can just keep

Mending nets at a yellowtail farm, Japan.

rotating through them.

But could there be a flaw in this plan to manage the production of aquatic foods so that traditional wild sources are no longer a troublesome bottleneck? This is a controversial topic, with some experts arguing that wild fisheries are a write-off, that the sensible plan is to domesticate the oceans with new technologies like free-drifting fish pens that follow the ocean currents, or herding fish at sea. Research, this argument goes, will solve the problems. Nonsense, say others, offshore fish farming will never benefit the poor and it'll deplete the supply of the fish needed to feed the farmed species. Protected areas are the way to go, whole spreads of aquatic habitat set aside for natural rebuilding followed by controlled harvesting.

Discussion is fine, but the problem is, if you're a policy maker in a developing country and this is the kind of rhetoric you pick up, what relevance does it have for your local situation? Take the São Francisco River, with fisherman wondering if they should accept the offer of the Ministry of Agriculture and start a small tilapia farm.

What will be the result in ten, twenty years? Where does all this meticulous planning and control really lead?

What happened to Japanese salmon

The day after I found out that the thundering noise in my in-laws' house was only *otōsan's* heels, Hatsumi and I caught a flight to Hokkaido. I was glad to get out; Tokyo alarms and oppresses me, and the argument that you can get anything you want there, while true, carries less weight when you can't have it with fresh air or removed from the company of thirteen million other people. I've come to feel comfortable in my in-laws' neighbourhood; in nearby Kichi-joji I've found the good places for coffee and an upstairs used-CD shop where the staff recognizes me and I can buy long-out-of-print recordings I would never find anywhere else. But the rest of it is still anonymous grey canyons and neon, like the futuristic city of *Blade Runner*. The crows start to get to me, patrolling the parks and side streets, and as big as the ravens back home. So do the wacky fashions. On this visit, the teenage girls in Shibuya were all in platform boots, miniskirts, artificial tans and cowboy hats. They had blonde hair and white lipstick and they came at you in droves. They called it "folklore style."

But outside Tokyo, Japan alternates between the pastoral and the remarkably wild; Hokkaido, I was told, was the place to be close to nature and away from people. Good, because this was a holiday. When we left the ground at Haneda Airport I looked down at the wasteland of concrete with its expressways threading like eels between the skyscrapers poking into the yellowed air. I thought, what happens if something goes wrong? How does a species react when they've become adapted to living like *that?* What kind of survival skills do you learn by twiddling your thumbs into a cell phone? On the way to the airport, the Rainbow Bridge had been clogged in gloom, the Sumida River beneath it a mess of cruise ships and hydrofoils. Good riddance, I thought, and sat back, waiting for the mountains.

And mountains appeared, surprisingly quickly. One of the many delights of travelling in Japan is that so much of the natural beauty remains available to explore. The old Tokkaido Road that was the traditional trade route between Osaka and Tokyo is now an uninter- rupted carpet of concrete, and the traveller on the train between the two centres may have trouble telling where one ends and the other begins, but you have only to strike out to either side of this conur- bation to be rapidly in rural Japan. For the visitor, leaving the city is easy, by bus or even by rented car — easier than for the locals, who are forced to enter and leave all at the same time, creating traffic jams that stop anyone from going anywhere.

I sat back and watched the Japan Airlines safety cartoon, which seemed more about politeness than what to do in the event of an emergency. Kicking the seat in front of you, the cartoon suggested, or drinking too much were the things to focus on, not something mortifying like an engine flaming out. Coughing was out of the question. I have been on flights to Japan in which, when the announcement is made that window shades "might" be lowered to allow passengers to enjoy the movie, any sunlight is shut out in sec- onds. They can't get those windows down fast enough. From space, the aircraft must seem like a many-eyed monster screwing its eyes shut against the sun.

But the first mountains we saw from the plane weren't the ones in Hokkaido, the ones I had read about in the guidebooks, charted hiking trips in, planned a trip through to the barren peninsula of Shiretoko, a word that means "end of the world" in Ainu and that juts into the Sea of Okhotsk like a frozen finger. They were less than a half-hour out of Tokyo, hugging the southern coast of the island of Honshu, small enough to be forested to their tops and traced, almost at regular intervals, by rivers flowing down to the agricultural plains and out to sea. It was like looking down on a relief map.

"Look at that," I said, pulling my wife across me so that our noses were pressed side by side against the plastic window. "See those rivers down there? Dammed, every one of them." Two-thirds of the way down the mountain, each river bloated suddenly into a reser- voir, tiny by Brazilian standards but big enough to generate enough electricity to contribute to the grid. "See, some of them even have a couple of dams."

Hatsumi said, *"Mmmmm?"* a drawn-out and very Japanese humming back in the throat, rising at the end, that signifies roughly equal parts polite interest and bafflement. Like most Japanese — like most people everywhere — she had never worried about where the energy she used every day to dry her hair and heat her bath was coming from. But I had read about how, in Japan, almost every river had been dammed or diked or otherwise drastically modified by the addition of tonnes of concrete. Now here it was unfolding down there beneath us. River after river. And on some of them, I told her, pushing on the top of her head so she was looking almost straight down, you could even see little clusters of buildings at the mouth.

"Hatcheries," I said. "See them?" She shook her head and wriggled free. "It's the Japanese way," I went on, breaking all the rules of cross-cultural marriage. Planning and more planning, a national network of salmon hatcheries that's practically eliminated the act of natural spawning in rivers too wrecked to support it anymore. Ocean ranching on a grand scale, unfolding before our eyes. What did she think of *that?*

"You promised this would be a holiday," Hatsumi said, and went back to watching the news on the TV monitor. The heads of the suited news anchors bobbed up and down as the news was punctuated by ritualistic bows. With the sound off, they looked to me just like the animated figures who had told us not to cough during the flight.

We landed at Kushiro airport and picked up our rental car, a microscopic Toyota with four-wheel drive that would surely be of no utility whatsoever in Japan's cities but made sense in snowy Hokkaido. Insurance for the car, as far as I could make out, was a million dollars a day. I decided to forget about all those zeros, put the car in gear and headed off down the wrong side of the road. An alarm chimed discreetly and the navigation system said something in Japanese.

The voice was lifelike, quite irritating. I already had a perfectly good paper map and I didn't see what this person could add to it, but I was a *gaijin* and would have to go along. I imagined a severe, middle-aged woman in an expensive linen suit. She wore purple eyeshadow and a choker of Murano glass beads from her last trip to Italy. Her hair was waved and dyed a fashionable light brown and she was seated behind some kind of console or desk so that I couldn't see her legs, but I thought they would be sturdy, hardly tapering toward the ankle, and pressed together tightly at the knees. Age was creeping up on her; there was a faint crease at mid-thigh, beneath the linen skirt, where her girdle ended. We named her Miss Tojo.

"She says you're on the wrong side," said my wife.

"I knew that," I said, moving over. There wasn't anybody else on the road anyway. We entered Kushiro, a seaport about a half-day's drive from Shiretoko, looking for something to eat before the drive into the national park. There were two vile-smelling pulp mills on the outskirts and I imagined Miss Tojo pressing a lavender-scented handkerchief to her nose as we drove past. The place felt like Nanaimo, "seven potatoes." I was right at home. A string of malls lined a nondescript main drag. Inside one of them, a search for a better road map led us to a "bookstore" the size of a football field. It sold mostly glossy magazines. I wondered how many British Columbia trees had gone into all these magazines, with their half-life of a week or less. An entire aisle was devoted to pornography. Many of the covers showed schoolgirls peeking out from beneath sweating salarymen. I could tell they were schoolgirls because they still had their socks and loafers on. We never found a road map. I could hear Miss Tojo saying, "I told you so," and making a *tsk*-ing sound.

Farther on, the malls were replaced by faded, older buildings. Hatsumi had programmed the navigation system for city hall, on the theory that a good noodle shop would be easy to find in the centre of town. I concentrated on driving. Stay on the left, I told myself, stay on the left. With every wrong turn, Tojo-san became increasingly annoyed. "Excuse me," she would say, and I imagined her lips tightly compressed. "You have taken a wrong turn." She might as well have added, "again." She would compute furiously, her screen going momentarily black. Then, "You must take the next right, followed by the second left. Please." And a graphic of the new route would appear on the screen.

Miss Tojo, I began to feel, was all about control. She would have a dog, a tiny white one, and when she took it for a walk in the park she would put on special gloves to pick up its poop and wipe its butt. She was one of these people who could never be satisfied, and there was a whiff of the authoritarian about her that made me wonder just how deeply her connections ran. Did she have friends in the Motor Vehicle Department? In the Police Department? I headed back the way we had come, stopping at a near-invisible booth, a six-seater counter operated by two women who served us extra-spicy miso *ramen* while Miss Tojo fussed with her lipstick and fumed.

Back on the road, Hokkaido felt even more like home. The countryside reminded me of the Cowichan Valley on southern Vancouver Island, the alders just in leaf and most of the land in pasture. The occasional barn reflected in a mirror-like rice field brought me back to reality. There were feed lots and off-sales of cheese and yogurt, tractors in front yards and used auto parts for sale. Barns rotted and sagged just like in Canada; the only difference was that the faded signs were in Japanese. The bleached hulk of a school bus abandoned in a field could have been an hour from my own home. So could the problems Hokkaido was having working all this agricultural land: many of the local kids didn't want to stay home and farm, so governments had to import gangs of young people from Tokyo to spend the summer in Hokkaido. One group heading for the big city, the other escaping it.

We sped toward mountains capped in snow, the smell of manure flowing in the window. I shoved all thoughts of fisheries to the back of my mind, into the storage area crowded already with dogs left in kennels and teenagers left in the house, and concentrated on having a holiday. When we passed a small hydro substation near a river and its eerie electrical buzz momentarily filled the car, I swallowed hard and kept quiet. The roads were immaculate and Miss Tojo was silent too, having told us clearly how to get to the fishing village of Rausu and then retired to file her nails or make life miserable for someone else. She didn't even seem to care that I was exceeding the absurd fifty-kilometre-an-hour speed limit, but with the roads nearly empty even a martinet could afford to let down her hair. Between Kushiro and Rausu she spoke only once.

"Excuse me," she said. "You have been driving for two hours. Please take a short rest." But her heart wasn't really in it, and she didn't insist. We did make a brief stop to see the fumaroles at Mt. Izoi, where foul-smelling earth-farts find their way out through yellow-encrusted blowholes in the rock. A group of teenagers scampered beyond the warning signs to get closer to the steaming orifices, another sign that Hokkaido was a place where rules could be bent.

On the way to Rausu we stopped at the famous "Five Lakes." Four of them were closed due to bears, so we had to share the pathway around the remaining lake with several tour buses full of middle-aged Japanese couples experiencing the outdoors. I made some unappreciated remarks about crowds and humiliated Hatsumi by breaking away from the designated trail and striking out across boggy moors until I reached the cliffs of the Shiretoko Peninsula. This was more like it. There was nobody out here and I could stare inland at snow-streaked peaks, or out to sea, toward Russia, imagining the ocean filled with creaking ice. I sat on a rounded rock and ate some cookies from a crinkly plastic package and when I finally got up to go I realized I wasn't really alone after all. Something was sticking up out of the moor. I dug it out: a camera, an ancient Canon, something from the fifties, before the flood of Japanese electronics hit North America. Its leather case was green with lichen but I got it open in a little shower of dirt and insects and, inside, the focus ring still turned. There was probably a fifty-year-old roll of film in there, snapshots of a family seated right on this rock, the kids eating *onigiri* while the mother warned them to stay away from the cliffs. I buttoned the camera and its memories back up, shoved it back in its hole in the turf and pulled out my own Canon to take a picture: a kind of ancestor worship. Then I trudged back to apologize.

Rausu and Utoro are fishing towns on opposite sides of the Shiretoko Peninsula and I liked them both immediately, even if they forced me to abandon my promise to stop thinking about fish. Rausu is on the milder side of the peninsula, if you can call it that, while Utoro sits on the Sea of Okhotsk. To get from one to the other you go up and over the wrinkled spine of the peninsula — the Shiretoko Pass. We hit Rausu in late afternoon, first encountering the usual armoured shoreline made from a jumble of concrete jumping jacks the size of our Toyota, an appalling and familiar sight throughout Japan that may be great at stopping erosion but looks terrible and over which you clamber at your peril. The things are called *teterapodo* — tetrapods — and they exist to sell concrete. Hatsumi told me she once fell between two of these giant castings

More concrete: the end of the river at Rausu, Japan.

and had to wait for hours to be rescued.

"You were, what, a little kid?" I asked. "On a school outing?" Field trips are big in Japan. Hatsumi looked at me strangely.

"It was a . . . party."

"Ummmmm?" I said.

Somebody in Rausu was obviously making a good living from the sea; the fishermen's houses on the approach to town were spacious and well-kept, their ample yards strung with nets and floats, with racks of split fish and squid facing the sun. I saw a clutch of tiny bars and restaurants, then a majestic bronze statue of a bear as we crossed the river and headed uphill toward the mountains. I tried not to notice that the river itself seemed to be missing its estuary and had been neatly paved for the last kilometre or so (more concrete), like the aqueducts that run through Los Angeles. Outside Rausu we said good evening to Miss Tojo and poked around the residential area until we came across a bed and breakfast, the Minshuku Ishibashi ("stone bridge," obviously a name from pre-

concrete days), and stopped long enough to negotiate a room and
get the once-over from the jolly, hare-lipped landlady. Her *minshuku*
was in the foothills, across from an abandoned-looking wooden
building straddling a stream. We took her suggestion and set off up
the hill for the local *rotemburo.*

Rotemburo are natural, open-air hot springs. This one's excellent,
the landlady said: undeveloped, no tourists, only the locals. The perfect thing after a day's driving. There was a small parking lot with
a scattering of cars, and a path through the woods toward some
steam. Just before we reached the source of the steam the path
diverged: men's and women's. I would be on my own. The women's
path led to a small changing hut; the men's ran directly to their
pool, which was shielded by a low cedar fence. Another woman
joined us and we dithered together, trying to figure out the local
protocol. Definitely not coed, but that wasn't too surprising, as few
rotemburo are anymore.

The women's pool must have been inside their little shed, perhaps to keep them from noticing that the men's pool was so much
grander than theirs. If Hatsumi wanted a soak she would have to put
on a suit and sit in the shed; I would have to go down the path to
the men's pool, introduce myself to the lads already lying in it and
enjoying a well-deserved respite from the labours of the day, strip
and hop in. I cursed myself for having abandoned my Japanese lessons and four heads suddenly appeared above the cedar fence, each
one glossy and steaming. They turned our way. We retreated.

"Apparently," I said, "there's a bath house behind the *minshuku.*"

So we went there instead, sitting in a slippery cedar tub until our
skin wrinkled, and if I hadn't decided to take a walk before dinner,
I would have been able to keep my mind in the emptied condition
that forty-five minutes in a small sulphurous room produces. Just a
stroll across the street to investigate the odd-looking building I'd
noticed when we arrived, back in a moment. The place was barn-
like but with no evidence of agriculture, and curiously open to
anyone who wanted to poke their head in. I crossed the road,
rubber-legged from the hot tub, and nipped in the open door, and
if I hadn't taken a moment to let my eyes accommodate to the
sudden gloom I would have stepped straight into two metres of

water. My eyes adjusted, wooden catwalks emerged, and I realized I had walked into a salmon hatchery.

Back home in Canada we use raised fibreglass tanks for the baby salmon, so only the most determined or inebriated visitor could actually blunder into one, but the contents of these Japanese tanks were the same: thousands of five-centimetre torpedoes I could see clearly once I got to my knees. They were circling, growing, waiting for the day they would be released into the ocean, and all in a building sitting wide open by the side of the road in rural Hokkaido. The Japanese commitment to salmon hatcheries was something every fisheries biologist hears about; well, here it was. I went back to the *minshuku* where Hatsumi and two other guests were sitting down to dinner at a long picnic-style table that was already nearly obscured by food. I had to tell someone.

"That building over there." The dining room had windows along one side, so you couldn't miss the hatchery from where we sat. "You'll never guess what that is."

Hatsumi glared at me. "You're late." Being late for dinner in Japan is simply not done. The landlady was in full swing, racing out from the kitchen with platters and talking rapidly. She seemed to be laughing. "It's about the *rotemburo*," Hatsumi whispered. "I told her you chickened out." The landlady set half a king crab in front of me, its pink legs clenched, and said something pointed. The room erupted and she swept back into the kitchen for more. Hatsumi leaned over.

"She says, what's wrong with you, you have two dicks or something? Eat your crab." I was glad Miss Tojo didn't have to listen to any of this.

Dinner was complicated and lengthy, the kind of free-for-all where the last swallow of fish soup is an effort of will. There was much talk, and much beer, and slurps and smacks and whistles as crab legs were sucked clean. The high point for me was the crab roe, crunchy and membranous and swimming in purple ink. I asked where it came from and was shown an unmarked pickle jar. Don't tell anyone, said the landlady, who had just finished explaining that she was a confirmed environmentalist. The roe was illegal.

After dinner we walked down into the village and found a tiny upstairs bar, one of those eclectic places that tell you far more about the owner than they would themselves. These little bars are one of the more interesting ways Japanese people reveal themselves; they come in all kinds of weird themes, like plastic toys from the fifties or Yomiura Giants baseball cards. This one was hung with fishing nets, so dense they trapped the cigarette smoke like low-lying fog. The owner was a middle-aged woman in a sixties bouffant hairdo. She brought us our beers and then, unexpectedly, a plate of freshly steamed scallops. The only other patron was a man in his fifties, nattily dressed in designer jeans and a golf shirt. He was doing very nicely in *konbu,* he told us.

"Fishing here, not so good," he said in English. "So now I am kelp farmer." Another technology off the shelf. He canoodled with the owner, she connected the karaoke machine and they both added to the fug of cigarette smoke. I sensed a long night. Before we finally left, she charged us for the scallops — another illusion shattered.

I slept badly, my stomach grumbling. My dreams were filled with shoals of salmon fry. Miss Tojo appeared among them, in a white bathing cap and a matronly black one-piece, kicking rhythmically. She seemed right at home. In the morning the landlady took pity on me, silently removing the Japanese breakfast of fish broth, rice and raw egg and bringing me back to life with homemade yogurt, sliced fresh strawberries and thick-cut white toast. I left for Utoro with the address of one of the landlady's relatives in Port Alberni and the feeling that Rausu had a lot more beneath its sleepy surface than my little scratch had uncovered.

Utoro was bigger, with a sizable deepwater port that was also home to a bewildering variety of smaller vessels. I could have spent a week getting to know what they all did; some were obviously trollers or longliners, not that different from the boats back home, but others were clearly for servicing the aquaculture operations in the bays, broad, shallow-draft scows that grazed through the forest of buoys, turning on a dime. Utoro port was punctured by mountain tops that poked up through the water right at the mouth to the harbour, so startlingly green we might have been in Bali, not next door to Russia.

The entrance to the town was a two-kilometre stretch of fishermen's houses fronting the ocean, and I could see the fishermen of Utoro were doing well too. There were buoys just offshore, a whole city of them, with a single open boat making its way slowly from one to the next. The man in Rausu had mentioned kelp farming, maybe that's what was going on out there. I made a note to ask more questions about how it was done. But an image was already in my mind: beneath the bobbing markers, a labyrinth of kelp fronds filling the bay, tethered in orderly racks like an underwater Tokyo, streaming with the tide while patient gardeners in black neoprene worked their way along the avenues. I reminded myself that, apart from the little slip with the hatchery the night before, this was still a holiday.

But Hokkaido was turning out to be the wrong place to come to get away from fisheries. Just getting to Utoro from Rausu was an obstacle course. The only way I could have avoided seeing the massive salmon hatchery on the northern coast of the peninsula would have been to shut my eyes on the drive down the Utoro side of the mountains, and Miss Tojo would never have permitted that. So there it was, far below, a complex of buildings strung along the last kilometre of the Iwaobetsu River. There weren't any spawning salmon that would get past *that* obstruction. And we couldn't seem to escape it either because, sure enough, the tour boat we boarded the next day for a cruise up the coast went right past the hatchery, cutting close to the estuary. Except that the estuary was gone, replaced by a concrete raceway and a cluster of buildings. It still smelled like home to the salmon, so up they would blunder, swimming stupidly into the arms of rubber-booted members of the Utoro Fisheries Cooperative. There must have been fish coming in, because a half-dozen bears were hanging around the river mouth like scalpers at a concert, waiting to pick off the unwary.

And back at the dock, fisheries intruded again. I love docks, their smells of creosote and engine oil and bird shit, their creak and flex under your feet, the heat that comes up through their weathered wood in summer, and the underwater gardens just centimetres from the surface if you drop to your stomach and look down. I did that in Utoro, while the rest of the passengers fussed over a dog sitting patiently at the top of the ramp, a short-haired Shiba-cross with a long, intelligent face someone had doctored by inking in an attractive set of eyebrows.

Down where I was, the water column was dense with tiny fish, each a few centimetres long, the schools of them marked by the occasional flash of silver. They were salmon fry, and there were many more of them in the shallows in front of the hotel we had found not far from the town. Probably chum salmon, the species that has been virtually taken over by the Japanese hatchery system, probably released just days ago from a hatchery like the one in Rausu.

At dinner that night, the owner of the hotel confirmed it. He was tall and good-looking with a deep, news-reader's voice, something of a Renaissance man. Maybe Hokkaido attracted people like this. The hotel was a second career, after twelve years of consulting in Yokohama. His wife did the cooking while he entertained the guests, and in his off-hours he suited up and lowered himself into the Sea of Okhotsk outside his front door, looking for specimens of interesting marine life that he sent by express post to a global network of scientific clients. He brought us *miso* soup and *uni* we spooned straight from the cracked spiny shells and, in between, a pickle jar.

Hotel owners in Hokkaido were always holding up pickle jars, it seemed. This one appeared at first to contain only water, but on closer inspection I saw a delicate, transparent invertebrate the length of a fingernail, propelling itself around its little world with graceful beats of tiny wings. Collected early this morning, he told us, and headed tomorrow for a scientist in Germany on the lookout for new anticancer drugs. He hauled out a photo album and pointed to an expanse of white with a few splashes of red and yellow clustered near the centre. Another photograph showed him closer up, in a red dry suit and holding a chainsaw.

"Come back in the winter," he said. "Every day I go out there" — he waved at the sparkling blue ocean beyond the rock shelf — "with my Snow Cat and chainsaw. Dive under the ice, another world." It was Russian ice, to be strictly accurate, formed near the coast of Sakhalin Island to travel across the Sea of Okhotsk where it consolidated and grew into a white plain extending around both sides of the Shiretoko Peninsula and lasting from October to April.

About the millions of salmon fry in the bay, our host knew plenty. "Chum salmon," he confirmed through Hatsumi as the other guests passed the tiny transparent angel around. "Every Japanese river is either diked or dammed," he went on, surely an exaggeration but I knew what he was getting at. "And the hatcheries, they used to be run by cooperatives of fishermen but the return in the fishery isn't as good as it used to be, so the fishermen want out. They want the government to go back to operating them." That was depressingly familiar too, especially to someone from Canada; invest heavily in a technology that tacitly accepts the alteration of natural habitat, then throw up your hands when the whole system runs into the ground a few decades down the road.

The next morning we went out early onto the rocks in front of the hotel, taking advantage of a low tide. Rocks isn't the right term, because the entire bay was a shelf of conglomerate, level and near-uniform, as though the place had been painstakingly filled with concrete. They can't do that, I thought ... can they? A few spikes of rock had detached themselves from the volcanic headland but the rest was flat, with a surface wrinkled like the skin of a cooled pudding and pockmarked with craters that, at low tide, had become aquaria. And, when I got down and looked, each aquarium was stocked with a few hatchery salmon fry, trapped until the ocean came back and reclaimed them.

A MONUMENT TO HATCHERIES

Our final stop before heading back to Tokyo was the Salmon Park in Shibetsu, at the base of the Shiretoko Mountains. Miss Tojo knew all about the place; it was in her database, although I couldn't imagine it was the sort she herself would choose. I saw her instead in Florence, Prague, even Banff — anywhere the shopping was good. But she took us to the Salmon Park without comment, although we could easily have found it ourselves, a futuristic cluster of buildings fronting a full-scale hatchery you could look down on from an observation tower. Attractive displays commemorated the six species of salmon native to Japan and instructed us about their life cycles.

The migration display at the Salmon Museum, Hokkaido.

Two things stuck in my mind.

The first was a three-dimensional display featuring the quadrant of the globe that contains the Japanese archipelago. The land masses were green and mountainous and the oceans a shiny, curving royal blue. Fine white lines traced the routes the salmon took on their oceanic peregrinations, extending from the coasts of Japan far into the Pacific and the seas of Japan and Okhotsk. The lines reminded me of the route maps in the back pages of in-flight magazines, except that the flights the salmon took were always return: east past Kamchatka, swim around in the North Pacific for a few years chasing shrimp and squid, then head home to spawn when the ticking of the biological clock becomes impossible to ignore. The entire display took up most of one gallery, and its very size and sweep made it impossible not to marvel at the glorious implausibility of the salmon's lifestyle. Creatures with brains the size of a pea unerringly finding their way along routes covering thousands of kilometres and returning to the very point of departure — not close, mind you, but within *metres* — this was something even Miss Tojo would have to take off her hat to (I imagined a woolen Borsalino, by Dolce and Gabbana). No wonder those Japanese scientists had designed their robot boat.

The best part, though, was the little red button at waist level because, when you pushed it, the islands of Japan lit up like Christmas. Tiny red lights, none of them inland, all of them crowded around the coastline. Shikoku had only a few, Honshu had more, a dozen or so, but the island of Hokkaido was ablaze. I had to push the button several times to count all seventy of them and I know there must be more because I doubt the one I almost fell into in Rausu was on the map. The little lights are hatcheries. And these days, every salmon heading out on one of those transoceanic loops starts and ends at one of those blinking red lights. And every red light marks the spot where there used to be a functioning, salmon-producing river, and now there's a concrete chute. That's planning for you.

The second best thing about the Hokkaido Salmon Park was an unintended reminder of how an idea — say, turning every estuary into a hatchery — gets around in the world of science. On my way out, I peeked inside an obscure little room tucked away next to the elevator that took you up onto the roof to look down on the hatchery. Inside were a number of disconnected exhibits related to salmon: bits of fishing gear, facsimiles of research articles in Japanese, faded colour posters of the six species of Pacific salmon. It looked like all the stuff nobody could figure out what to do with but nobody wanted to throw away, like the clutter under my in-laws' kitchen table in Tokyo. I scooted through the room and was about to go back to Miss Tojo when I did a double take.

Those posters looked familiar. They were standard issue Government of Canada fisheries department posters, unchanged in thirty years, the kind I could point out to you in any fisheries office, school or museum in western Canada. There were "our" chinook, coho, sockeye salmon, each species rendered brilliantly in paintings Harry Heine made thirty years ago and reproduced so many times that the number of trees felled in order to print them could probably have sheltered many a salmon stream.

Any traveller knows the twinge of stumbling across some artifact from home. It's a pleasant feeling, although it can be odd (my two weirdest: the Stanley Cup final watched in a seedy hotel in Caracas, and a pair of wooden snowshoes in the window of a Belo Horizonte junk store). On that level, seeing the poster of "Canadian" salmon in Japan was hardly an experience worth relating. A Japanese tourist in Canada would get a bigger jolt, certainly a laugh, from the salmon museum at Hell's Gate on the Fraser River, where the name is carefully spelled out in Japanese for the benefit of the thousands of tourists that pass through: *Health Gate.* On another level, though, that poster was a clear message about the reach of science, about how technical developments can spread around the world. That poster represented the process by which countries like Brazil could end up managing their fisheries to a model created halfway around the globe. How does that process really work?

Somebody brought those posters from Canada in a suitcase — a Japanese scientist returning from a technical tour in Canada, or a Canadian embarking on one in Japan. Probably there was an international conference somewhere in the itinerary, which meant plenty of mixing, formal exchanges of theories and findings, beers shared and ideas and experiences bandied back and forth. Those beers, the delivery of scientific papers, the visits to research labs and the ritual exchange of posters and bottles of sake, all represent the extraordinary power and reach of the scientific method. Scientists compete with each other, but they're still a gigantic old boys' club, with chapters everywhere.

The science club works by connectedness and consensus, and while it's fundamentally unbiased, even anarchic, once its findings have become enshrined in public policy it produces dogmas as pervasive as religion and as hard to bring down as sumo wrestlers. The sumo wrestler in this case is hatcheries, which embody both the power science has given us to control fish reproduction (an achievement, make no mistake) and the unspoken agreement that this ability to artificially inseminate fish and release their babies into the wild *should* be used to keep the supply of fish coming.

The use of hatcheries to bolster fish populations is politely called "enhancement" and, in terms of sheer production numbers, it's been a spectacular success. Hatcheries are great at turning out fish, and enhancement has often ended up meaning replacement. But here is the problem: like any other successful idea, this one is getting exported. It's the product of decades of research and development, applied over even more decades by all of the salmon-loving nations of the world, who happen also to be the richest and the best positioned to export their success stories to developing countries in the form of "aid."

Japan is one of the most advanced fishing nations in the world, perhaps *the* most advanced. So is my own country, Canada; the two countries have a long history of cooperation in fisheries science. Both, not surprisingly, have been generous donors when it comes to "helping" less developed nations solve their own fisheries problems. Sometimes this works; other times it's a case of the blind, if not actually *leading* the blind, then stumbling together, hand in hand, toward the same precipice. The irony is that hatcheries are catching on in the developing nations at just the time when they're becoming a bit tawdry at home.

Hatcheries are under fire in North America now, for a multitude of sins that took decades to creep out from under their rocks, but how do you tell that to someone in the Philippines who thinks hatcheries and enhancement are the answers to the fishing industry's prayers? How do you tell it to a fishermen's cooperative in Fiji that wants to stock the ocean with sea bass fingerlings? Do these people know that the only people still pushing salmon hatcheries in North America are commercial and sport fishermen desperate to find a technical work-around for conservation restrictions on their take ("Just let us fish, we'll build a hatchery to make up for it!")? If you're a small-time fisherman on the São Francisco River and the Ministry of Agriculture paints a glowing picture of the benefits of enhancing your reservoir or wetland with hatchery fish, maybe even planting some in a cage in the river and abandoning fishing altogether, are you really going to argue? *It worked in Japan, didn't it?*

Back in the parking lot, our car was baking under an unusually warm spring sun. We were lucky, Hokkaido wasn't supposed to be so balmy in May. A school bus released some small human fry, who raced for the ice cream concession before being reined in by their teachers and formed into an orderly line for an hour's indoctrination about fish. Canadian kids are the same, an entire generation raised to believe that salmon come from hatcheries. We opened the windows and Hatsumi set the navigation system for Monbetsu Airport and I thought, It's all because of you, Miss Tojo. You and your friends sitting down to dinner in your favourite sushi restaurant, with your Prada handbags slung over the backs of the chairs in so-safe Tokyo, mellow jazz in the background and a little personalized box of restaurant matches for your cigarettes after the side of mackerel the waiter seared at your table, just so, using a hand-held blowtorch. You're the ones who demand the freshest yellowtail and that wedge of fatty tuna belly so smooth it melts against your palate. What do you care if the fish spent its last moments pinioned by rubber-coated Mexicans with knives and icepicks?

But it's not fair to pick on Miss Tojo because she's selfish, or on the Japanese because they like seafood and would rather die than question the status quo. Japan's the easy target because they've taken the technology the farthest, that's all. We're all guilty. The next time I go to Japan, I'll drink *otōsan's shochu* and eat his *maguro* tuna until I drop. In the developed world we *all* export technology and we never question too closely whether the example we're setting is a healthy one to follow. In the developed world, we all think we know where we're going. We're all Miss Tojos.

CHAPTER 10

LEADING THE BLIND

Everything you ever wanted to know about

fisheries management

That's not how we do it in Brazil: two cold Brazilians exchanging fish stories with a Canadian fisheries officer.

WE IN THE "DEVELOPED WORLD" THINK WE KNOW where we're going, and there's always been a natural tendency to invite less developed countries along for the ride. The first step is often the technical tour. I've perpetrated a number of these, and I'll describe a couple in a minute. If you're not careful, a technical tour becomes a junket; sometimes, though, given the right mix of people and a little luck, it can change lives. The usual result is somewhere in the middle, a case of the blind leading the blind. Especially if your subject is fisheries management.

I once went to a concert in a small church, a warm summer evening kind of thing. Four trumpeters played, unfortunately at the same time. The piece was a Bach partita, but the result wasn't Bach. The result wasn't even music. The volume and the cat-swinging closeness of the walls took Bach's impeccably ordered scheme and turned it into a fantastic barrage of harmonics, rings, howls. I leaned forward in my pew, rested my forehead on the Book of Common Prayer and slid my fingers up over my ears. Sense into nonsense: that concert could be a metaphor for what happens when a roomful of otherwise rational people discusses fisheries management.

So it was with Miguel Ribon and three other Brazilians one late summer day in Vancouver. Miguel drummed his fingers on the table and kept looking around, at the door. It was close to noon, he probably needed a smoke — there hadn't been a break since the meeting started at nine, not even for coffee. As though that weren't uncivilized enough, lunch was bound to be a disappointment. Beans and rice are hard to find in downtown Vancouver. I knew this but Miguel didn't. It looked like a long day.

I like Miguel, who has the biodiversity portfolio at the State Ministry of Forests in Minas Gerais, where the São Francisco River arises. He's one of those big men who seem oddly shapeless, whom you can only imagine in rumpled, forgettable clothes. Unlike many Brazilians, he wears his heart on his sleeve. Polite, affable, charming, garrulous — you can say that of most Brazilians, but Miguel was a little different, a little edgy. You knew where you stood with Miguel. Sitting next to him, Roberto Messias was more polished and more typical. I wasn't at all sure where I stood with Roberto. He was tough, though, showing no sign of having gone sleepless for two days when his flight to the tiny coastal village of Tofino was cancelled and his trip from Brazil degenerated into a marathon of ferries and buses. Roberto was the most senior person in the room, right at the top of the State Ministry of Environment, and clothes obviously mattered to *him;* he was wearing designer jeans, polished loafers and a snappy yellow shirt with a Ministry logo embroidered on the pocket. Roberto wore the same shirt every day of the tour to Canada that I'd helped organize; I wondered if he had a suitcase full of them.

Beside my left ear, Carlos Bernardo kept up a steady stream of asides. Carlos was a biologist who knew the fish in the São Francisco inside out. So did Norberto dos Santos at the other end of the table, but from another viewpoint entirely. Short, black, a wiry, fit-looking sixty with kinky grey hair and half-glasses, Norberto had been fishing the stretch of the São Francisco downstream of the Três Marias dam for forty years. It hadn't left him any time to learn English like the others, but we'd brought him to Canada anyway because if there's one thing fisheries managers have finally learned to do it's to include the fishermen in their meetings. And if you drag a bunch of Brazilian bureaucrats and biologists all the way up to Canada to talk fisheries, you'd better bring a fisherman along. Anyway, everybody knew Norberto. He was no fool, and a little thing like a lack of English didn't stop him. Right now he was watching Gary Logan, our Canadian Department of Fisheries host for the morning, as though he understood every word.

"Look," said Gary, throwing out his arms against a picture-postcard view of Vancouver harbour. "You need three things to manage a fishery, okay? You have to know how many fish are out there in the first place, so, stock assessment." He held up a thick finger. "Two, you gotta know how many you're catching. That's monitoring, right? And three" — up went the third finger — "you gotta enforce your rules. Observers on the boats, cops on the water, whatever it takes." The three fingers made a fist and his hand dropped to the table. Outside, the floatplane from Victoria skittered to a landing in the harbour, and I imagined the passengers hunched over, clutching their briefcases, peering out the portholes at the high rises that go right down to the water now. Most of them would be government bureaucrats with reservations on the five PM return flight. Behind the float plane, the Mexican three-master *Cuahautemoc* glided incongruously out under full sail. It was Tall Ships weekend, mid-July, last year's snow cap long gone from the North Shore mountains overlooking the city, the view from the Department of Fisheries and Oceans boardroom slowly filling in with construction cranes and half-built high-rises.

Gary Logan is stocky and pugnacious-looking, a long-time DFO biologist with a grey buzz cut, a boxer's nose and a no-bullshit manner that has helped him through countless encounters with disgruntled fishermen and made him the perfect choice to give the visiting Brazilians a rundown on the Canadian style of fisheries management. He seemed to have been chewing the same piece of gum since our meeting started. Many of Gary's sentences end with, "Okay?" and the people he's talking to usually get addressed as "buddy" or "guy." Once, when I ran into him, he was wearing a complicated splint on one arm, an articulated plastic contraption that looked part medical device and part hockey armour.

"Should I ask?" I said.

"Don't," said Gary.

And he was right about management. Who could argue with something so fundamental? Assess, monitor, enforce. If you accept that a fishery can actually be managed (and most people seem to, since we claim to be able to manage anything these days, even time) then counting and policing make perfect sense. I allowed myself a cautious pat on the back. It *was* right to have brought these long-suffering Brazilians on this dog and pony show; they *could* learn something about fisheries from us. It's all reducible to a few simple concepts, whether you're on the Fraser or the São Francisco — both rivers have fish that need to be kept track of if we want to keep catching them. It made sense to Gary and it made sense to me.

But what would it mean to the Brazilians? Counting fish means knowing where they are and being able to sample them reliably. How do you count a moving target whose multiple life stages, egg to adult, might as well be different species, they behave so differently? A target that's to all intents and purposes *invisible?*

Miguel seemed to have forgotten his longed-for cigarette and his empty stomach. He leaned forward and said in a gravelly smoker's voice, "But our rivers are different. You can't do this kind of stock assessment." Beside me, Carlos Bernardo grunted. Gary stopped chewing his gum and stared hard at Miguel. He had obviously been in enough encounters like this to know when someone meant what they said. Can't meant can't; *why* didn't matter. Everybody in the room knew enough fish biology to realize that counting methods which worked for salmon — a single, economically dominant species mainly fished in the ocean — weren't necessarily going to work on two dozen different tropical species in a big, muddy river. He started chewing again. I thought he might be chewing a little faster now.

"But catch surveys, you can do those, okay? Monitor what the fishermen take and calculate from that?"

Someone translated for Norberto the fisherman, who got a little smile on his face. "And how do you do these catch surveys?" asked Roberto, politely.

"Lots of ways. Landing reports, every time a vessel comes in. Annual reports, every licence holder has to submit one." I could immediately see how impossible that was on the São Francisco, where most of the fish were caught by small-scale operators, a couple of guys in a canoe with a cast net or hand-held hook and line, their catch landed anywhere along the shore and vanishing into unregulated roadside markets.

"Observers too," Gary went on, warming to the subject. "The

observer program is huge for the department. We put observers right on the boats. Contract the whole thing out. Now we're even pushing the industry to install video cameras on deck, as a condition of their fishing licence." He was on a roll.

"Video cameras." Miguel was shaking his head.

Gary chewed harder. "Hey, for us it's a good investment. An observer costs the licence holder four-fifty a day, and you can install a camera for ten grand. Pays for itself in a season."

Video cameras; ten grand; around the table, Brazilian eyebrows went up. What was the possible relevance to a flotilla of mom-and-pop operations strung out along two thousand kilometres of tropical river? There had to be a link, or we'd dragged these people halfway around the world to listen politely to a lot of information they couldn't use. Maybe that pat on the back had been premature. More development dollars squandered. I pulled at my chin.

Roberto said, "Well, your salmon, that's a big industry. But for us . . ." He held out one index finger and waggled it, side to side: wrong turn, don't go there. You see the gesture all the time in Brazil, along with my favourite, a repeated flick of the wrist, like shaking water off, that snaps the thumb and third finger together, *click click click*. It means "faster!" and I would like to have used it now, if I were able to do it.

"Salmon?" said Gary. "I'm not talking salmon here. The observers, the cameras, this is all on the trawlers, the groundfish boats. Biggest sector of the seafood business here, unless you count farmed fish and we won't even go there, okay?" He did some more chewing while the unimportance of the wild salmon fishery sank in. Visitors to the west coast of North America are still blitzed with images of salmon: smoked, in elegant cedar boxes in the airport, leaping at the end of an angler's thrumming line, tours of "Fishermen's Wharves." Because the reality isn't tourist-friendly. A slimy black cod nosing around in the abyssal darkness? A million Pacific hake, all fins and lips, dumped on deck in a slithering mountain en route to being puréed into *surimi?* Salmon are still sexier.

Gary looked like a boxer sensing an opening. "Pound for pound, the real money around here is in rockfish." Rockfish: spiny, multi-hued, deep-dwelling and bug-eyed; some people call them snapper.

Rockfish was Gary's specialty. He sat back in his chair and linked his arms behind his head.

I asked, "How long is the coastline in BC? How many kilometres are we talking about where people are catching rockfish?"

"Thousands," said Gary. We settled on a rough estimate of a couple of lengths of the São Francisco River, if you pulled all the coves and reefs of rockfish habitat into a straight line. "It's mostly for the Asian market, especially around Vancouver. Twenty-three bucks a kilo around Christmas time, live. All hook and line, mostly small boats working on order from the wholesalers. Big fishery, okay?"

Someone made a quick calculation and a translation — twenty-three dollars a kilo was nearly fifty reais — definitely a lot. And the small boats part, using hook and line: that was getting a bit more Brazilian, even if the fish themselves definitely weren't. Rockfish live deep, in a vertical world, hanging out along undersea cliffs like underwater alpinists. When they get dragged to the surface their eyes pop out and their swim bladders blow up like balloons, protruding from their mouths like air bags after a crash. The Asian market only wants fish of just the right size, the couple of kilos that looks right on the plate. Those ones get their swim bladders swiftly punctured and they can survive in a live tank for another ten days or so, plenty of time to be trucked to a city wholesaler, change hands a few times, be sold, killed and cooked.

The ones that don't measure up are just as unlucky, except the end comes sooner. They get yanked off the hook and tossed back, where their blown-up swim bladders prevent them from ever swimming again. The practice is called high-grading and it's like bycatch of unwanted species, a punishing biological tax that can account for half the total catch. High-grading means that, if you're a rockfish, you either spend your last moments backed into the corner of a glass aquarium while the chef's dip-net closes in, or staring at the unfamiliar sky before the seal shoots up to get you.

"I'll tell you a story about this fishery," Gary said. He had everyone's attention now. "December 24th, 2004. Just before Christmas, okay? We seized a semi-trailer full of live rockfish, none of it reported. We'd been following the feeder boats from the Queen Charlottes for weeks. The guy, the one behind the operation, was a

Vancouver guy, convictions for drugs already, a known criminal. It's still in court."

This was getting more relevant by the minute. Fish caught illegally in Brazilian rivers are constantly finding their way to markets in the big cities. Expense-account bureaucrats in Brasilia and high rollers in São Paulo will pay well for something like *pirarucu* (which is endangered in many places) and even *surubim,* the big catfish that make up so much of the São Francisco fishery. I didn't expect the Brazilians to be too shocked by the criminal element in the rockfish story either, because a lot of the illegal fish in Brazil zips across the country in the trunks of taxis or under false floors in small pickup trucks. You don't mess with these guys; the few fisheries officers that exist in Brazil wear guns for a reason.

"What happened to all the fish?" Miguel asked. I could see him calculating how many rockfish would fit in a semi-trailer. Gary looked pained.

"Destroyed," he answered.

"Oh my God." In Brazil, seized fish get given away to schools and food banks. Maybe not so much seized as should be, but at least it doesn't get dumped.

"I know," said Gary. "It's bad. We used to give them away, but the lawyers told us to stop. The liability would kill us now."

Someone asked about fines, another sore point for Brazilian fishermen who've historically been hit with penalties that could amount to a year's wages for a gear infraction they consider ridiculously unjust in the first place. Gary told them how fines paid to the Justice Department went back to the DFO to fund conservation programs; how a couple of recreational rockfish fishermen had recently been fined $14,000; how fisheries officers were going after a rash of illegal sales that inevitably follow reduction of the allowable catch. Fewer fish, fewer opportunities to fish legally, more illegal fishing. Gary and the Brazilians went back and forth.

Enforcement of fishing regulations is something most people don't think much about, probably don't even know goes on. It's a tall order anywhere, and in Brazil it's nearly impossible — not so much for lack of legislation, but because of huge gaps in training and capacity. Decent, fair enforcement means a lot more than a gun

Canadian fisheries officer Herb Redekopp and visiting Brazilians.

and a pair of shiny boots. The kind of training a fish cop gets in Canada is not only unheard of in Brazil, it's undreamed of.

Policing is a curious offshoot of the business of catching and selling fish, and it has its own character. Think about it: major investment in boats and gear, a time-sensitive product worth anywhere from a healthy wage to a killing, and a theatre of operations that keeps bobbing up and down. Fisheries enforcement is unique. If you're a farmer, is a uniformed officer likely to pop up on the back forty and ask for a look at the potatoes in the back of the truck? Do loggers have to worry about armed conservation officers materializing suddenly from behind a tree? The only time that might happen is if the logging was happening too close to a river, in which case the police would be fish cops anyway.

I looked out the boardroom window and let my mind wander. More float planes came and went; more container ships crept into the harbour. The enforcement vessels wouldn't come in here; around the corner, near the Fraser mouth was where you'd find them. We sat inside and talked about Brazilian fisheries but, outside, salmon season was starting.

FISH POLICE; FISH POLITICS

In the fall of 2005, I rode with a federal fisheries officer on patrol in the lower reaches of the Fraser River. The big, rigid-hulled Zodiac flew over the roiling water past container ports and potash dumps and all the other signs of a heavily industrialized river mouth. The twin Yamaha 225s were so loud you had to wear communications headphones. It was an odd sensation: even though we were screaming along at thirty-five knots, the voice in the big green earphones was relaxed and conversational, as though we were facing each other over beers in a quiet pub. The cop driving the boat was telling me the story about a raid on some Canadian crab fishermen poaching in U.S. waters.

"They're Vietnamese, most of these fishermen. South Vietnamese, North Vietnamese, Chinese Vietnamese. They don't even speak to each other. It's a border fishery, plenty of temptation to bend the rules." I knew the place. The forty-ninth parallel that divides the U.S. from Canada runs just south of the Fraser River's mouth, and it's a rich environment for crabs. I imagined the American crabs and the Canadian, sidling through the eelgrass, feelers waving in the current, crossing from one nation to the other.

"First there's the Coast Guard cutter, it comes steaming straight at them, and at the last moment out from behind it come four rigid-hulled inflatables like this one, fanning out around the crab boats." The boats try to scatter; the air fills with illegal crab pots heaved overboard.

"No dice, suddenly there's a Black Hawk helicopter coming in low, it starts flying a tight circle around the crab boats, pinning them down." The cop turned around and grinned at me, zooming his free hand down.

"They tossed one of the guys in jail, didn't hear his case for a week, and in the end he lost his boat, which is quite an investment."

"And then?"

"Business as usual. It takes a lot to faze some of those guys."

I peeled the headphones off and watched a couple of late-season Native gillnetters go by. Of *course* the Vietnamese crabbers weren't fazed. In Vietnam, fisheries enforcement is a bit different. In the Tonle Sap, the huge lake on the Mekong River that borders Vietnam and Cambodia, there are lucrative fisheries for a dozen freshwater species and they're controlled by local warlords, who have created an allocation system based not on all-stakeholder meetings in panelled boardrooms, but on mutual understanding about just what will happen if the fishermen working for one boss stray onto somebody else's territory. A bullet to the head is what would happen. A few helicopters and a week in a Seattle jail doesn't even compare.

Every fishery has its own politics. Violence happens, and not just on the Mekong. That summer, even as we cruised one section of the lower Fraser, disturbing images of young fishermen from the Cheam Indian Band were appearing in the newspapers as parties squared off over a sockeye fishery thirty kilometres upstream, a fishery the Department of Fisheries refused to open. Out on the river, T-shirts were pulled up over noses; enforcement vessels were rammed. And, for show, the spokesmen for the rival groups enacted a Mexican standoff that reminded me of the final scene in the classic spaghetti western *The Good, the Bad and the Ugly.*

In the movie, three gunfighters arrange themselves in a graveyard at equidistant points, and the camera cuts from one to the next as their hands hover over their pistols. Their eyes, squinting against the sun and shot in pore-rendering close-up, dart this way and that. If I shoot first, whom do I aim for? Will the third guy get me in the back?

That summer, the commercial fishermen (Lee Van Cleef, in the role of Angel Eyes) pleaded publicly for a sockeye opening. Privately, they had no intention of taking the government up on it. That's because even a small opening would trigger an aboriginal fishery (Eli Wallach, as Tuco) not just for aboriginal consumption but for sales, something the commercial guys vehemently opposed. And the commercial guys thought they knew why the natives would want such a fishery — they were convinced it would allow them to open cold storage vaults full of fish caught for "food use only" and sell them. Over our dead bodies, thought the commercial fishermen, those fish can rot and we'll cut off our own noses to make sure they do. The Department of Fisheries enforcement cops like the one taking me up the river now (Clint Eastwood, the Man With No Name), just watched and waited: they couldn't move until someone else did.

Hands hovered over pistols, eyes darted. Tuco complained about racist fishery policies, Angel Eyes complained about poaching. In the end, the fishery was never opened that year. The Man With No Name never fired a shot. Deciding who was "right" was nearly impossible. The cop driving the boat probably knew more secrets than anyone. I put the communications headphones back on and asked the obvious. "So who's telling the truth?"

The cop laughed. "They're all fishermen, right?"

I nodded, waiting for the punchline.

"So they *must* be liars."

"You like this job?" I asked.

"Love it," the cop said. "I can make more money working down-town, though." Downtown was in DFO headquarters, where the boardrooms were. "But I prefer it out here. Better toys." He slapped the steering wheel.

"Ever pull out any bodies?"

"Not me, but another officer based upstream at Chilliwack pulled out thirty-seven in his career. People kill themselves way up in the Fraser Canyon and the bodies drift down here. I did my swift-water training up there, in Dead Man's Pool. One time I grabbed a log going around and around but it turned out to be a cow. All the hair had rotted off."

A voice broke in on Channel 16, the marine communication band: "Um, advising you the vessel Double Time is attempting to tie off nets across the entire river up by New Westminster."

"I'll stop by and have a chat with him." The cop hung up the mike and turned to me. "Third-party report. Somebody thinks they saw something. The boats are supposed to leave a third of the river open, by law. Let's talk to this guy first" — he pointed at a vessel across the river — "then head up to New West." He bent the Zodiac into a sharp turn. The Yamahas howled and we skittered sideways and then bore down on the other boat.

We came alongside and cut the engines. Two native men were shaking chum salmon out of a gillnet being wound slowly onto the drum at the stern. The chum were already battered-looking and they came in along with bits of wood and weed; no bodies that I could see. Everybody knew everybody. The two Indians kept yanking fish out of the net and the cop bantered with them and looked around. From where we were floating I could see a plastic tub full of salmon roe on top of the wheelhouse. After a few minutes we took off again, burbling slowly away until we were clear of the orange buoys that marked the net, then opening the throttles so that the boat lifted and levelled and flew on upriver toward New Westminster. I put on the headphones and asked about the roe I'd seen. The cop grinned.

"Oh, probably sell it to a tackle store for bait, but you won't see any money change hands."

"What about the chum?"

"They can catch it for food. Legal fishery, no problem. If a little gets smoked and sold, hey, what can you do?"

"And that's all they can catch?"

"Chum and hatchery coho, yeah. The net's supposed to be selective so none of the sockeye get caught by mistake. Sockeye have to be released." He made a wry face. "Who in his right mind would release a dead fish?"

This was the Mexican standoff in action. The cop may have preferred it to life in an office, but that's where our Brazilians were now. In the grandeur of the fifteenth-floor boardroom I sat back and watched Gary and the others go at it, the Brazilians' questions coming fast, and tried to ignore my own rumbling stomach. I felt relieved. Yes, the Brazilian and Canadian species were different, and their habitats, and the methods used to catch them. But the principles for all fisheries really were the same: know what you have, count what's being caught, catch and punish the cheaters. If you don't do these things you lose the fishery and you could even lose the whole resource.

The meeting broke up finally, manly embraces all around. Gary stood and stretched, putting his arms behind him and easing his shoulders: he had another meeting in half an hour. He looked like a boxer between rounds. The rest of us straggled into noon-hour Vancouver to look for rice and beans. Once out of the confining boardroom the Brazilians strung out behind me as I searched for a restaurant. Several of them disappeared, pausing to light cigarettes or peer into shop windows. I found the rice they craved, at a Flaming Wok in a subterranean food court, and waited for everyone to catch up. There weren't any beans.

How much are we catching?

"Diverse fishery, diverse river," an international fisheries expert intoned to me once in a bar, draining his mug of beer and setting it down on the wet Molson's coaster. He was a little guy from the north of England, one of those characters who have spent their professional lives moving from one continent to another, one project to another, chasing the funding, weathering the politics, dropping relationships and families along the way and constructing new ones out of the local materials, learning so much they can't even get out a sentence without making me feel as ignorant as an undergraduate. Diverse fishery, diverse river; it sounded good, what did I know?

Dogmas are addictive. They should be kicked every decade or so, if only to make room for the next one. Take the most-quoted phrase in news reports on global fisheries: "Seventy per cent of all fisheries are overextended or at their limits." FAO came up with this in the mid-1990s, and the media have been using it ever since. A friend of mine in FAO winces whenever he sees the statement; he also places the fingers of one hand against his temple and looks down at the floor, as though a particularly toxic gas bubble were working its way through him.

"Fisheries statistics," he groans, "there's an oxymoron for you. We get our statistics, catch and species and weight and all that, from the FAO field offices. Do you have any *idea* what that means in a developing country? Especially one without much infrastructure? The field office could be the corner of someone's desk, and he might not have been paid for months. He might not even exist! Best case, he does exist and even compiles reports — but they might never reach us, or be so incomplete as to be useless. I mean, how can we expect a guy to report catch statistics when nobody records catch in the first place? Make it up?" He grimaces again. "No, wait, better not answer that. Somebody from China might hear you." China's reported fishery statistics were constantly being flagged as inflated up until 1998; thereafter, the country announced a "zero-growth" policy, where marine and inland catches obediently levelled off. Even with this "adjustment," China still hauls in a quarter of all catches from inland waters.

Until recently, most biologists and managers went along with the idea that river fisheries are basically different from the big ocean fisheries that dredge species indiscriminately off the bottom or cordon off vast areas with so many nets the fish haven't got a chance. Most of the big fishing-collapse stories have been about this kind of take-no-prisoners industrial fishing in the ocean. River fisheries, the argument has always gone, are mostly small-scale subsistence affairs, villagers in canoes with cast nets out to support their families. The few industrial fisheries, like the ones in the Amazon, were considered anomalies.

But even if it's still largely true that river fisheries are small-scale and spread out, that also means they're virtually impossible to monitor, so nobody really knows how much fish is being taken out. And some of these river fisheries are actually capable of driving down stocks just as relentlessly as the more notorious ocean fisheries. The Mekong is the best example. In Cambodia, each small boat can take out five hundred kilograms in a week, and the bigger boats with bag nets can remove nine tonnes in a day. In the lower river, huge species like the Mekong giant catfish, which can grow up to three metres long, have been fished to near-extinction.

In the Tonle Sap, the largest lake in Southeast Asia and one of the most productive inland fishing grounds in the world, the number of fishermen has steadily increased, with each taking less and less. Not surprisingly, the average size of the fish being caught is dropping too, so that most of it is now small, low-value perch instead of the big migratory species of the past. To call these fisheries "small-scale" is absurd.

One belief still holds: inland waters are at the mercy of horrible things happening to habitat, which is where everyone still points the finger. It's just that this is not the whole picture, and the impact of fishing in inland waters may not be so small after all; there are multiple villains. Nevertheless, over-fishing will continue, because most of the inland fisheries are open access, and fishing is often the lowest occupation on the economic totem pole. If poor people can't fish, what are they supposed to do?

If countries really want to get a handle on over-fishing in their rivers and lakes, even if it's only to decide whether it's important at all in comparison to all the other things that afflict inland waters, then they're going to have to know how many fish are being caught. It's called stock assessment. All the developed countries do it, but mention it in a tropical river basin and hackles rise. Fishermen especially don't like the idea; it's convenient to keep blaming someone else for the declines (stick to the dogma), and stock assessment looks an awful lot like an excuse to strengthen enforcement. And it's hard to do when fish are landed at hundreds of different locations up and down what amounts to thousand-kilometre coastlines. It's not like driving down to the nearest ocean port and collecting the data from all the fish boats that have to come there; small-scale river fishermen can unload anywhere.

There's also the problem of the "shifting baseline syndrome." If you're measuring numbers of fish in a fishery, what is "normal"? If somebody says catches are "down," what's the point of reference? Biologists get their first feel for the state of their field when they start their careers; whatever the stocks are then, that's their baseline. When I started out, there were still plenty of salmon going up the Fraser River; nobody worried much, we were too busy catching them. Stock strength in 1980 was my baseline. Historical abundances many times greater might be found in the old reports, and there is always an old fisherman or native elder with stories about the old days when, heck, nobody even needed a boat, all you had to do was walk out on the backs of all those fish and grab a couple, eh? Which might very well be true, but it's not what each subsequent generation grows up with. Little by little, we become accustomed to less. For most people it's just nostalgia, but for a fisheries manager it's a trap, one of those subtle biases built into resource management.

But stock assessment still has to be done. I once endured a two-day workshop on stock assessment for the São Francisco, sitting with my chin on my hands in a stifling meeting room attached to a *pousada* near the river. It wasn't anything like the *pousada* on the beach in Ilheus with its oiled-wood furniture and Gauguin prints and the smell of the surf outside. This place was stifling and enclosed. From time to time, someone would get up and un-tack the blankets draping one bank of windows and move them to the next, trying to keep the place cool and make the images projected onto the wall a little less washed-out. Two horses looked on from a field outside, from the shade of an avocado tree whose fruit hung low and solid, like waxy green testicles.

Fishermen were scattered throughout the room, seated according to their political affiliations to this colony or that. You had to feel for them, because everyone else was there too: biologists with their graphs and charts, government management in their monogrammed shirts. Two members of the Environmental Police sweated steadily into their brown polyester uniforms, their combat boots, their holsters; these guys were on the front line, but the chances of their knowing the regulations they were supposed to enforce were slim. You have to remember, this is a country where it's technically illegal to hang your elbow out the car window.

Some of the fishermen were just trying to make a living; others were middlemen running orders of fresh fish up to Brasilia in the trunks of taxis. As far as they were concerned, stock assessment was a ganging up, a way to gather data to be used against them. They listened patiently to experts from the Amazon and the Paraná on the benefits of recording catches, along with hard proof that catch was dropping even as effort increased (something called catch per unit effort, or CPUE, that every fisherman knows in his bones).

One obvious and major problem was geographic. Even on the stretch of the São Francisco between, say, the dam at Três Marias and the next one at Sobradinho, a distance of more than five hundred kilometres and only a quarter of the whole river, there are major differences not only in the availability of fish but also in the way the local resource is managed. Três Marias is comparatively rich in fish; Pirapora, two hours down the road, is poorer. Both are fractious. The next community, though, a place called Ibiai, is something of a poster town for fisheries harmony, with an innovative arrangement for dividing sales between individual fishermen. The heads of the *colonias* for all three communities were at the meeting, and all three had a different take on where stock assessment would fit into their own little world. Lumping them all together as "fishermen" didn't make sense.

The message wasn't getting through. On the second day, the head of the fishing federation was conspicuously absent. He'd been there the day before, a serious, dark man with a head like a coffee bean, taking notes one laborious letter at a time, his arm curled around the paper, going back to fill in the accents. "There are plenty of fish," he'd said then, "it's just that they're getting smaller." You couldn't find a better description of a fishery in deep trouble. A Canadian expert on stock assessment, Neil Schubert, was listened to politely as he soldiered through his presentation in Portuguese. Brazilians speechified. Knuckles were cracked. The horses switched their tails. There was a muffled whump from somewhere outside and the room emptied briefly but it was only someone's windshield exploding in the parking lot: a cigarette lighter left on the dashboard. We all filed back inside, where a bee the size of my thumb was blundering repeatedly into a window, the only evidence I'd seen in two days of someone actually trying to get *into* the gathering.

I doodled on the back of the program, working on my wrap-up speech, fretting about the best way to say "we appreciate the effort you have all made to make this meeting a success" while avoiding irony. My mind, already dangerously misfiring because of the heat, began to wander. Was it possible to calculate the amount of effort expended to bring all these people to this room to discuss a subject most of them wanted to avoid? Nature tends toward entropy, toward disorganization and chaos, but here we were, tables, projectors, pencils, sweating policemen, the swimming pool outside, even the blankets on the windows a challenge to the Second Law of Thermodynamics. Extra energy was keeping it all from flying apart, mostly human energy, but even the horses outside were doing their part. If they didn't stop cropping at that dried grass they would cease to exist. I imagined them crumpling to the ground, dissolving, devoured, vanished. Considered in these terms, our little meeting was an achievement.

At the end, a senior bureaucrat from the national fisheries agency, a bombastic nonentity with a red, granular nose and tenuous connections to President Lula, stood up and announced that his agency would host a follow-up meeting. No expense would be spared to ensure that the momentum created here would be transferred, indeed multiplied a hundredfold, to his administration. A comprehensive program of stock assessment was as good as done. He sat down and went back to whispering in Neil's ear about, I found out later, all the important people he knew. Unfortunately, Brazil creates a new national fisheries agency every year or so. The current model — its people, its policies, its desks and its projects — flies off in all directions like interstellar debris. The guy with the red nose probably has a different job by now, and fisheries in the São Francisco still aren't being counted.

NEW KIDS ON THE BLOCK: THE SPORTIES

You generalize about fisheries at your peril, but there are some trends that can't be ignored. One of these trends is who, demographically speaking, is actually catching the fish. Some significant shifts have happened here, mostly in the last twenty years or so, to the point where large fisheries that were once dominated by one group have been taken over almost completely by another. The São Francisco and the Fraser are examples.

Not everyone who fishes the São Francisco does it clinging by his toenails to a rock in the rapids, or from a crudely built wooden canoe with hand-carved paddles like the one Amadeu was working on when I visited him and his silent wife in Barra de Guaicui. More and more of them do it for fun, a motive that creates a divide between two very different groups of people.

When you're on foot in the middle of a boiling river and suddenly have to share your point of contact with a thrashing fish the size of a small child, is this *fun?* From the promenade along the river in Pirapora I can look across the rapids to Buritizeiro. The fishermen working the rocks with their cast nets are nothing more than the tiny spots of colour of their shorts: a red one, with a small explosion of silver as the net flies out; a dark blue one inching imperceptibly toward the shore, like a blip on a radar screen. I try to imagine a man's emotions when the net suddenly dances in his hands and a golden *dourado* or spotted *surubim* breaks the surface. Fear perhaps, or nothing at all, just a desperate ballet involving a blizzard of automatic calculations of weight and balance and friction that, if successful, result in gaining the shore with your life and a small portion of your livelihood. Professional fishermen will always tell you they like fishing, but it's the independence, the besting of the elements, the wrestling of a living from nature that they refer to. I have rarely heard one say that actually catching a fish is fun.

I have never fished professionally, and ceased years ago to try to catch anything for sport (or "recreation," the now-favoured term that eliminates any suggestion of the fisherman having fun at the expense of the fish). I used to enjoy sitting in my boat in a kelp bed, jigging cheap lures between the waving stalks and pulling up the occasional rockfish or snapper. It was nice to be out on the water, but I lost more lures than I caught fish, and there were too many people with the same idea. The number of rockfish dwindled to almost nothing; Gary Logan and his colleagues wisely restricted the fishery; I sold my boat. I am not a suitable spokesman for angling.

Nevertheless, the amount of sport fishing going on can tell you a lot about the future of a fishery. I went with Norberto dos Santos to observe the sport fishery on the São Francisco near Três Marias, an area where it's already well developed and where new, comfortable *pousadas* are springing up along the banks to cater to visitors from other parts of the country. Norberto was in his element, released from the polyester slacks and street shoes he'd brought to Vancouver. Now he was back in shorts, T-shirt and flip-flops, picking his way down steps cut into the bank below his house and headquarters. A fifteen-horsepower Mercury outboard was balanced on one sixty-year-old shoulder and the other hand held a cellphone to his ear. His little fleet was bobbing below, six aluminum twelve-footers jostling in the current beneath the green canopy.

Norberto used to fish for his family, but this fleet is his future.

Norberto dos Santos multi-tasking.

The sport fishermen drive slowly down the dirt road, looking for the faded hand-lettered sign that says "Sr. Norberto," and he and his guides take the visitors out. The visitors, *graças a Deus,* keep coming. Every day the Mercury outboards go up and down the steps because they're too big an investment to leave alone all night; every few years, a new boat gets added to the fleet. I wonder if Norberto is the modern equivalent of the *barqueiro* Richard Burton describes, the men that made up his team of boatmen for the trip downriver. "All men here are more or less amphibious; the canoe, they say, is their horse" — yes, that fits, and so does Burton's remark that the best of the *barqueiros* eschews foul language and drink, is quiet and intelligent, has a penchant for what Burton calls "mild chaffing" and a delight in song. Definitely Norberto.

The bows of the boats nosed together into the bank, like cows around a trough. I followed Norberto into one of them, grabbing a warm aluminum gunwale in each hand and vaulting quickly into the centre of the boat before it had a chance to rock. Norberto finished connecting the fuel line and tossed me a faded life jacket, stiff and stained and missing its buckles. I have a complicated relationship with life jackets, especially as a passenger; they remind me I'm not in control. I saw myself bobbing ridiculously, flailing in and out of ill-fitting straps, my feet kicking feebly at the catfish cruising disinterestedly below and no more visible from the shore than an orange rind. I doubted if the vest Norberto had handed me would even float, and when I shrugged it on, the collar stood like an Elizabethan ruff.

Norberto swung the boat backward in a tight circle, kicked the motor into forward and we took off upstream, angling out from beneath the shadow of the trees into the centre of the channel. I took the life jacket off and sat on it. Almost directly overhead, the bridge to Três Marias clanked and rattled with the unceasing train of charcoal trucks bound for the smelters outside Belo Horizonte, and I thought back to the first time I had been on that bridge, how Barbara Johnsen had patiently explained things to me: trucks, *cerrado,* refineries, potholes. We accelerated past her shit grass, fed by the sewage of Três Marias.

At water level, the surface of a great river like the São Francisco is roiled and uneven, especially near the banks where water claws at the land as it rushes by. Rivers need to do that, pulling soil away here and depositing it there, redistributing solids and nutrients in a natural cycle that dams and their reservoirs dismantle. As the boat gathered speed, its metal hull rang against the creases and ridges of all that turbulence; I felt the sound as well as heard it, my feet vibrating with the thin metal skin of the boat. I remembered the first boat I was ever in, a one-cylinder putt-putt my father had rented for the afternoon at a small marina in British Columbia. None of us — my father, my brother, me — knew anything about fishing. I asked my father about the slapping sound under my feet. "Salmon," he said. "Salmon hitting the bottom of the boat." I had no idea what a salmon looked like. It could have been small and silvery, a shoal of them fluttering beneath our keel, or as big as my father, a leviathan sliding casually beneath us, ready to heave our little craft into the air with a lazy sweep of its tail. What they looked like didn't ultimately matter because we never caught anything. But I never forgot that sound.

Norberto opened the throttle and the boat rose to a plane, striking upstream. The slaps of imaginary fishes were replaced by an occasional *whump* as the boat surfed over the wrinkled surface of the river. I wedged my hat under the floorboards and followed Norberto's pointing finger to a break in the trees: a dock, two figures in lawn chairs, a glint of monofilament slanting into the current, a blue-and-yellow parasol brilliantly backlit by the morning sun. It was a sport-fishing *pousada*. We rounded a gentle curve in the river and there was another one, bigger this time, with boats like Norberto's tied to a wooden jetty. I wondered how many years until Norberto could afford to build his own little hotel. Farther back in the trees, I caught a glimpse of buildings and the flash of white as an incurious cow picked its way through the vegetation.

What looked like a forest was often only a narrow strip fronting the water; behind, the irrigated fields were getting bigger every year. The vegetation was changing; even Burton noted "second growth where magnificent forests had been." I picked out a *jatobá* tree, one of the few I could recognize. There are churches in Minas Gerais, dating from the sixteenth century, with altars made from *jatobá,* warm and massive and glinting across an entire wall. For me, the *jatobá* is a big, impressive tree glimpsed from an aluminum boat on the river, the name hastily harvested and jotted down before it disappears in our wake. Not for Burton, who managed to record its many names, its appearance, its uses. "It produces gum anime, a good pottery varnish, and a copal used by the Indians in making their labrets and other ornaments; the flowers are enjoyed by the deer, especially that called *mateiro,* and the long chestnut-coloured pods that strew the ground supply a flour of insipid taste, which serves, however, in times of famine."

Burton did this for everything: plant, animal, fish, rock, human. Reading *Explorations of the Highlands of the Brazil,* you stumble suddenly onto a half-page of detailed botanical and ethnographic descriptions of a dozen plants. The "poor man's quinine," for example, is a "tree with bitter bark and seed fruit" — and Burton then provides its eight local names and the reasons for each. Or the *carnauba* tree, the most valuable palm of the *sertão* (the backlands), still important in rural Bahian economies: the gum and *pulpo* are edible, ribs and fibres make fences and nets, and the leaves when dried exude the stuff which, mixed with tallow, forms carnauba wax. And finally, "The most beautiful growth is the *ipê amarello* . . . its trumpet-shaped blossoms, in tufts of yellow gold, would make the laburnum look dull and pale." He must have scribbled non-stop.

We continued downriver, passing the occasional chunk of dirty styrofoam that marked an illegally tethered hook. Every kilometre or so, a pair of powerlines spanned the river, hung with birds. Norberto cut in closer, extending a thumbs-up to a waving figure before bending our boat around toward the opposite bank, a place where the river seemed to curve suddenly and where the surface, even from this distance, looked roughened and uneasy. As we bore down, what looked like a bend resolved itself into a constriction, the banks closing down and the river crowding between them in glossy humps of rolling brown water.

Burton's descriptions of the many rapids of the São Francisco would fill an appendix, but luckily he didn't believe in such things, expecting the reader to put in the effort to absorb, *right now*, the fruits of his prodigious note-taking. Burton knows, and gives, all the names for all the varieties of waters to be found on the São Francisco, from "mere broken waters *(quebradas)*" to the long runs called *corredeiras* with their little steps or *corridas*. (The rapids at Pirapora are *corredeiras*.) In the lower third of the river, he and his men ran one rapid after another; my favourite has the wonderful name of Destaca Calções — loosen your breeches — which is, as Burton drily notes, "all that required be said of it." He must surely have encountered the one Norberto was now taking me through; it looked ancient and very permanent.

As we got within a hundred metres, Norberto slowed and headed for a pathway he must have known all his life: an uneasy ribbon of calm between the tops of two boulders that broke the water like the foreheads of river giants crouching beneath the surface. The current shook us violently as we shot through, Norberto flinging his arm out across the river. I followed his pointing finger and there was another boat, on the other side of the channel but pointing into the current. We cut behind them for a closer look: three men, one crouched over the outboard, scanning the confused

Sport fishermen in rapids near Três Marias.

river as he deftly kept the boat stationary in the eye of the mael-
strom. The other two were fishing.

That made three groups of fishermen in five minutes, all of them
amateurs, all of them spending money to drive here from Belo or
São Paulo, to stay at a *pousada,* to buy gear, even, like the ones parked
in the rapids hoping for a strike from a migrating *dourado or surubim,*
to hire a guide with a boat. Add up all this money, factor in the
employment created, and you get an argument; more than that, you
get a lobby in the enviable position of being run by people with dis-
posable income and time on their hands. We left the fishermen
bobbing in their tenuous equilibrium and Norberto continued
toward the bank, making for a tiny clearing I saw at the last moment
before he cut the engine and the bow ran into a tangle of branches.

Empty pop bottles washed down from Três Marias poked out
from between roots, and a fishing pole was jammed into the caked
mud. I ducked and grabbed at one of the branches as we came to a
stop and a *goiaba* — guava — swung down and tapped my cheek,

lumpy, cream-coloured, the size of an orange. I pulled off a dozen to share with Norberto. They were the perfect fruit, the ulcerous and unappetizing skin surprisingly delicate around the seedy pink interior, the whole thing edible in an instant. Norberto waved his away, already out of the boat and crouched over the fishing pole. He began to pull in the line.

I'd seen this before, in the Taquari River that flows through the huge wetland of the Pantanal in western Brazil. That pole had been wedged into a wooden dock and clamped by rocks big enough to suggest that whoever left it was either an optimist or knew a lot more about the local fish than I did. My companion on that trip was a Brazilian biologist named Celso with a dazzling grin and the unsettling habit of whacking me on the shoulder with a palm like a ham and exclaiming, "My big little friend!" Celso shrugged and said, "Somebody hoping for a *surubim* to come along tonight." The fact that fishing was closed for the reproductive season didn't seem to enter into it, and with one enforcement officer for the whole river the chances of getting caught were exactly zero. Even I, the *gringo,* knew that the one officer was days away, in another part of the Pantanal, worrying about more ambitious poachers and whether his budget would stretch to another tankful of gasoline.

But at least poachers didn't have a lobby, unless you counted intimidation. Norberto had the hook out of the water and showed it to me: small stuff, a wad of something jammed over a hook too small to hold anything serious. I sniffed the evil-looking brown ball. Peanut butter, definitely, but something else too. Brazilian fish, it seemed, like cheese as much as Brazilian humans. I struggled to remember the joke someone had told me, and together Norberto and I reconstructed it:

Genie (I imagine something bald, confident and affable, say the soccer star Ronaldo with his gap-toothed smile): "You have three wishes. Number one?"

Fortunate Brazilian (without hesitation): "Cheese."

Genie: "Of course. Number two?"

Brazilian: "Cheese also."

Genie: "Naturally. And the final wish?"

Brazilian (after some internal struggle): "Sex with a beautiful woman."

Genie (sternly): "Are you sure? Explain this choice."

Brazilian (hangdog): "I was too embarrassed to ask for more cheese."

Norberto tossed the line back in and we ate guavas together under the trees while a family of golden monkeys skipped between the branches above us like irritated chaperones. I watched a toucan glide through the understory. A turtle sunning itself on a log cautiously put out its scaly head. We talked about fishing on this stretch of the São Francisco: who caught what, and how, and why, and what would happen if the numbers of fish kept falling the way the available scientific evidence suggested they were. It was the halting, one brick at a time conversation of people who don't share a common first language or culture; my Portuguese, as always, ran like a twig navigating a thundering rapid: shooting forward here, yanked backward there, re-emerging downstream after a long disappearance. I wondered if Norberto knew anything about the sport fisherman's lobby in Canada, and decided to try to convey something of its character by describing Bob Chambers, a passionate advocate for sport fisheries on the lower Fraser River.

Bob is sad-looking, with limp greying hair and a wheedling voice. Bob slouches. He's retired, a useful qualification for being an advocate of sports fishing — although I was never sure from what. I imagine him at a desk, one of those synthetic-veneered ones for the lower ranks of bureaucrat, in a fabric-covered cubicle. There are portraits of Bob's family and Jesus thumb-tacked either side of a large calendar whose every month is illustrated by a grinning fisherman wrestling a sparkling salmon. I see Bob stamping papers, tax files perhaps, the top of his head gleaming under rows of harsh fluorescent lights, his mind far away on salmon and dreams of retirement. Occasionally he sends an e-mail. The ones I received from Bob always ended with a Biblical exhortation, a different one each time, so that a rambling account of the iniquities of some group he felt was stealing his allocation of salmon would be followed by CLEAVE TO THE ONE TRUE LORD or LOVE YOUR NEIGHBOUR. The capital letters reminded you who was on Bob's team.

None of this was translatable to Portuguese, or at least not my Portuguese. But the attitude surely was: there were fewer fish now, and it was only right that those fish go where they produced the biggest economic benefit. Bob is just a foot soldier in an army of sport fishing advocates, all of whom are stuck on the same doctrine. Gas, food, lodging, revenues from fishing licences, all the toys a guy needs to go fishing — the value of all these things gets toted up, packaged, hurled at the heads of managers and legislators as though an army of paperboys on bicycles were taking aim at a bleary-eyed man on his front porch. To people like Bob, conservation is of the Ducks Unlimited variety: the end product is something to kill. "We don't care if it's wild or if it came from a hatchery, whatever," Bob told me once. "We just want to fish."

You can't fairly compare the sport fishermen on the São Francisco, with their deck chairs and their little brown balls of peanut butter and cheese, with someone like Bob. But it's only a matter of time: the glossy magazines already exist, pulling people to the river, and up in the Amazon the industry is already well developed, with the Pantanal coming up fast behind. Real fishing lodges are being built up there, not the laid-back *pousadas* of the São Francisco but fly-in places with splashy advertising on the Internet and clients from Germany and Japan and the U.S., all dreaming about a fat peacock bass dancing at the end of their line.

But there *is* something that's comparable, and that's the disparity between the local, professional fishermen, the ones who fish for subsistence or as a small business, and the visitors who want their share of the resource. The amateurs fishing along the São Francisco may not be the high rollers that arrive on the Amazon by float plane and sit down to a gourmet meal every night, but in comparison with the local fishermen in their canoes, they might as well be. They have connections and clout, and when there's an almost total lack of data on how many fish are actually being caught (as there is on the São Francisco), then connections and clout end up tipping the balance. It's your word — your lobby — against theirs. In Bob's case, you also have to answer to the Lord.

I learned about the Brazilian sports fishing lobby seven hundred kilometres away from the São Francisco, in Brasilia. Brasilia is where the power is, where the deals are made; it's the kind of place that empties on Friday afternoons. The top bureaucrats in Brasilia wear dark suits and stylish spectacles and they leave their cell phones on. In the high places, the embassies and the offices of directors and deputy ministers, you can taste the power, sipping from mono-grammed china demitasse brought to the boardroom on a silver tray by gloved servants. Down below, ordinary people dash across Brasilia's sticky avenues because the planners forgot about pedestri-ans. Brasilia is a city for cars and expense accounts.

Years ago, a man called Washington leaned close in a deafening bar in Brasilia and said something that made me dislike him. Washington is high up in the Environment Division of a hydroelec-tric company. I try to imagine what his job is like, fielding accusations from fishermen that the turbines are annihilating tons of fish, balancing competing requests for this release of water or that minimum flow, being responsive always to the American parent who snapped up the utility in the mid-1990s. In Washington's office, people wear suits and ties and there are storerooms stacked with easy-on-the-eyes brochures about the company's environmental performance, books on local fish fauna the company has subsidized, T-shirts and mugs and lapel pins for conferences and environmental research stations and community awareness days. Washington's office is one of those places you leave heavier than you arrive.

He was a big man, smooth and soft-looking, sweating heavily into his suit jacket. "Forget it," Washington shouted in my ear. "The professional fishery on the São Francisco, it's over." He sat back and took another drink of *cachaça* while the beautiful couples of Brasilia flowed around us, circulating between tables, embracing, dancing. Tomorrow was Sunday; maybe some of them would take mass in the Catedral Metropolitana, beneath Oscar Niemeyer's suspended bronze angels.

"From now on, sports fishing. That's the way it has to be." But no one really knew how many fish were left, and Washington's ear was simply more accessible to the lobby with the better connections. Rights, wrongs, even facts become irrelevant in such a situation, and the loudest voice is the only one heard; worse, those with the loud-est voices forget, gradually and with the encouragement of their

No town for pedestrians: Brasilia.

nodding audiences, that there was ever any other way.

That's the worst of it: the latecomers have the shortest memories. Having given Norberto Bob Chambers, I tried next to tell him about the stand-off between fishing lobbies on the Fraser River — something that seemed to happen every time there weren't enough salmon to go around. As you drive east from Vancouver through the valley of the Fraser, the mountains converge over the fields of late-summer corn until, after nearly two hours of thundering four-lane traffic, you come to the town of Hope. It's a dividing place, with baffling road signs that alarm the first-time traveller who now has to choose between three highways, all of them leading more or less through the mountains and on toward the Rockies. Hope is a turning point for the salmon too, the last stretch of easy sailing before they head northeast up the Fraser canyon, beyond which they face their own choices of highways in the form of tributaries that extend a thousand kilometres into the interior plateau. The salmon, of course, need no roadsigns; they veer unfailingly up the path they took four years earlier in the other direction, following cues so subtle — smell, taste, maybe even the magnetic pull of the earth's core — that science still scratches its head.

The sockeye skirmishes often take place just before Hope, along a stretch where the Fraser broadens and slows after racing through the mountains. It's an idyllic place, the first chance the river has had to rest, dropping the sand and silt collected from the interior so that the banks are sandy and smooth. A hundred kilometres farther west and it will have braided and split before expiring in a plume of silt rolling into the Strait of Georgia, but here, near Hope, it's accessible for the first time to fishermen in boats. The area has been a traditional Indian fishing ground for generations.

When test fisheries in the approaches to the river show too few migrating sockeye to allow the commercial fleet the openings it wants, lobbies clash, especially if the native fishery, specifically the non-commercial harvest for food and ceremonial purposes, is allowed to go on. Accusations fly: illegal native fishing, illegal sales, government favouritism and lack of enforcement, reverse discrimination against law-abiding fishermen whose livelihoods were petering out. Second-class citizens, the sport fishermen call themselves.

What was the difference between this idyllic stretch of the Fraser and Norberto's river? Whose rights mattered most? How did we reach the point where a comfortable middle-class group interested in nothing more than a good time in their off-hours could dictate to a culture who were surviving off the fish long before Europeans even knew the place existed? By the same process, I suppose, that a Brazilian hydroelectric manager can pronounce one fishery dead and another taking its place: ignorance, bluster, self-interest.

Norberto and I talked under the trees while the monkeys kept an eye on us and the sun tracked higher in the sky. By noon it was finding its way through the canopy and our resting place was getting uncomfortable. I grabbed a few more *goiaba* and we pushed off, back upriver. The outboard had to work hard against the current and we fell silent for the long trip home. I wanted to tell Norberto about my encounter with Washington in Brasilia, but my Portuguese was played out and the story was too complicated. It was too bad, because there had been a sequel.

Years later, after Washington told me the little guys were finished in the São Francisco, the same man sat in my kitchen. Another group of Brazilians was visiting Canada, and we were working our way through chicken and steak and pineapple grilled outside and sliced into bite-sized shreds, Brazilian style. An old guitar retrieved from under a bed was on Washington's knee, and he was singing. The songs were mostly his own, in a wavering tenor at odds with his bulk. One song followed another, about his home, his wife, his family, about Velho Chico, the São Francisco. Wherever this was coming from was surely a different, deeper place than the source of his pronouncement that had so offended me before. A tear rolled down one cheek and he sang on, unabashed, and I was reminded that things on the São Francisco were never as simple as they seemed. Washington was human after all. I decided to forgive him. Bob Chambers would have been proud of me. A few months later, I heard that he had revised his views on the sharing of fish in the São Francisco, that there was room for subsistence fishermen after all. Maybe, in Washington's case, the technical tour actually worked.

PART III

Velho Chico

CHAPTER 11

THE RIVER'S LUNGS

Tropical rivers need wetlands

Someday, all this will be yours: in front of tilapia cages in the Marituba wetland, Alagoas, Brazil.

THE IDEAL RIVER JOURNEY BEGINS at the source and ends at the ocean. Rivers are nature's narrative; all you have to do is follow along. It's that comparison with the ocean again: if you push off from the seashore in your boat, wind or tide will probably bring you back in soon enough. You have to work hard to get anywhere on the ocean, striking offshore until the currents or the trade winds carry you away. Getting to those currents is a preamble, like wading through page after page of a long-winded preface until the story finally starts. It's different with rivers, where the plot unspools right from the first page. Push off from the edge and you're gone.

With some rivers, though, especially rivers as big as the São Francisco, "just following along" is easier said than done. The story takes twists and turns. Even in Richard Burton's time, before the dams obliterated great sections of free-flowing water and forever altered the flow of what remained, there were obstacles — rapids and gorges — that no boat of his time could negotiate. Burton's expedition had to put into the water below the rapids at Pirapora, entering the river a fifth of the way from its source by way of a turbulent trip down the Rio das Velhas from Belo Horizonte. And he took his boats out of the water at the cataract of Paulo Afonso, a hundred and fifty kilometres before the river reaches the Atlantic.

Going by road, even today, would be worse, especially in the section that runs through the interior of Bahia. Even on the map, the roads here look tentative, and many of them connect villages of just a few hundred people. Some of the paved ones are cross-hatched with the dreaded symbol for "precarious road." Depending on the season and the amount of money that's found its way to public works since the map was printed, you can expect holes the size of your car, and plenty of them. My map helpfully provides emergency telephone numbers next to each of these sections, although I'm not sure what would actually happen if I tried to use one (in Brazil, cell phone companies mark their territories with road signs, but there's often a no-man's land, which is where I would have a breakdown). For much of the stretch between Pirapora and Paulo Afonso, your car is a pinball, and the points where it actually touches the river are few. You would never even know, from glancing at the road map, that there was a major river running through it all; when you do locate the São Francisco it's distorted along its length by the seven dams, like a deflating party balloon. On the engraved map that accompanies Burton's book, the situation is reversed: the towns are fewer, there aren't any roads, and the river is unimpeded, a highway from start to finish.

A few people have even traced the river on foot. On October 4, 1992, the day of St. Francis, patron of the São Francisco River, four "pilgrims" including a certain Franciscan friar performed a symbolic baptism at the source of the river in Minas Gerais. They filled a bottle of water in Canastra, where the river begins, shouldered their packs and headed downstream on foot. It took a year to cover the three thousand kilometres to the Atlantic. As they progressed from mountains to *cerrado* to arid *caatinga* and finally to the sea, they talked, listened, took notes. They saw everything; they spoke to everyone. Their notes became a book and their remarkable journey launched a national campaign for revitalizing the river. If anyone could be said to have paid the price for learning the true story of this suffering river, it was these four people.

The friar was Brother Luíz Cappio. Frei Cappio is the one who has said, of wine made from grapes grown in irrigated plantations along the river, "The wine of the São Francisco is made from the blood of the people." Frei Cappio's journey in 1992 was only the beginning of his very public involvement with the São Francisco (we will hear more from him later) and it's a hard one to repeat. Unless you have a lot of time and a liking for unmarked roads and endless scrub, seeing the São Francisco means picking your points. Três Marias, where the dams begin, is an easy and logical entry and it was the first place I went, but to see how important the São Francisco was in the early years of colonial Brazil you have to go to the mouth, too.

The last hundred kilometres or so, from Paulo Afonso to the ocean, run along the border between the states of Alagoas and Sergipe, and are easily reachable from the city of Maceió just up the coast. The flight from São Paulo touches down in Salvador first and, once it takes off again for the half-hour hop north to Maceió, skims low over kilometres of beaches and sugar cane, as though gaining altitude just isn't worth the bother. The beaches where the first European navigators landed are on the left side of the plane and the reefs they had to pick their way through are on the right. The beaches look inviting from the air, as most beaches do, but as usual in Brazil the real picture is more complicated. Those Africans whose rusted fetters now hang on a restaurant wall in Ouro Preto — what would have been their first sight in Brazil? Maybe that bay right down there.

Fátima

Maceió is big, a million or so, and I was glad to be met at the airport by a friend of a friend. Such people are essential when travelling in Brazil, a country where networks are huge and elastic and always ready to accommodate another member. If you don't already know someone, you soon will. I still receive e-mails from a policeman to whom I gave a two-hundred kilometre 'lift' outside Pirapora. The poor guy flagged me down at a police checkpoint. By some insanity of job scheduling he had to leave his family behind in Pirapora and spend half of every week in Ouro Preto, eight hours away. He was a good guide, and so, it turned out, was the person who met me at the airport in Maceió.

Fátima was in her early fifties, solid but still beautiful, with long black hair tied in a ponytail and an Arabic hook to her nose. All I knew was that she was an expert on the local fish fauna. She picked me out unerringly from the arriving crowd.

"Brian!"

So much for blending in.

"Come on." I got a bear hug and Fátima led me to a small Fiat baking in the parking lot, sweeping CDs off the passenger seat. We found a hotel using the normal Brazilian method of driving from one to another, tramping up and down stairs until the right balance of price and amenities was established. The room I settled on in Maceió had bare white walls, but it was across from the beach. Fátima said something about an aged parent, hugged me again and promised to pick me up the next morning. I had the evening off.

The beach itself was what happens near a city: a row of hotels, most of them small like my own; a promenade of sorts, with kiosks selling food and beer and phone cards and surrounded by sticky yellow plastic chairs; some scruffy grass above the sand and a collection of sightseeing pirogues pulled up on rollers next to upended fibreglass fishing boats. By the time I got there in the late afternoon, most of the fishing boats were back on shore, and the small fish market right in front of my window seemed open. I walked across the street to have a closer look but saw only a three-legged cat nosing around and a man curled up asleep on the main display counter, his feet resting in a sorry pile of what looked like sardines. My knowledge of marine species has never been good.

Down at the water's edge, a man was bent and busy with wooden mallet and knife. I watched him dismantle a metre-wide ray, whacking the knife repeatedly to carve a channel around the guts and tail, lifting them out and tossing them into the sea, waving the remainder through the water and carrying it up to the market. With its guts gone the ray looked exactly like an old-fashioned life jacket. The man gave me the look reserved for *gringos* who have dropped directly from outer space. I didn't blame him; I felt spaced-out myself. I was tired; the flight from Canada had taken twenty-four hours, including the usual interminable wait in São Paulo, the connection north to Salvador and finally here. Now I was up to my ankles in blood-warm water and those ankles were the dead white of Canadian skin in November. I am sure I looked like a weirdo.

An empty plastic bottle came ashore, along with the smell of sewage. Up the coast from here, a few months ago, some biologists working on a sea turtle project had been robbed and killed. Probably chopped up with a machete, like the ray. I was seized briefly by the unreasoning fear of all travellers arriving for the first time in a strange place and I looked out past the mess of the shoreline to the reef a kilometre offshore, a silvery scribble beneath a line of black rainclouds. The water would be cleaner there; maybe sometime I would go.

But this was a river trip. Maceió was where Richard Burton had finished his own journey down the São Francisco. After the rafts were taken out and dismantled at the falls of Paulo Afonso, he seems to have been in a bit of a fog; who wouldn't, after such an experience? But he soon rallies with three therapeutic pages of botanical description, then trudges downstream to catch the steamer that takes him to the mouth, where Fátima would take me tomorrow. Mind you, Burton was connected: when he visited Maceió it was to see his friend the President of Alagoas. I had Fátima.

I bought a beer at the kiosk next to the market and crossed the street back to the hotel. As if to remind me why I was here, the public phone booth next to the kiosk wasn't in the shape of a dolphin or a bonito; it was, instead, a gigantic fibreglass catfish, a *surubim,* speckled and snouty. The *surubim,* says Burton, is often five feet long, weighing 128 pounds and yielding two kegs of oil. The people net it, and "the wild men shoot it with arrows. . . I have never tasted a finer fresh-water fish." In the nineteenth century they were even exported to Europe — "Brazilian cod," they were called.

Surubim are still the emblem of the river. I was, it seemed, in the right place after all. Back in my room I drank the beer and watched the colours on the reef fade and disappear. I knew I should get up and shower off the stink of airline travel but I could hear the water hissing along the beach and my eyelids were drooping. On the television behind me, small colourful figures scurried like ants after an invisible soccer ball. The last thing I heard before sinking into sleep was the drawn-out cry of *gooooooooooal . . .*

INVADERS

I awoke to bright sunlight and murmuring from the television, which still swarmed with tiny footballers. (In Brazil, soccer never sleeps.) Outside there was another sound, a kind of scraping; a man was swinging a straw broom at bottles and cigarette butts. Just up the street a flatbed truck dropped off another worker. He too wore a red peaked cap, shorts, white shirt with long red sleeves, and red knee socks. A half-dozen more, identically clad, were clustered on the truck waiting to be deposited farther along the strip. The operation reminded me of the maintenance trucks in Canada that creep along a road under repair, dropping off orange pylons. These human pylons looked pretty cheerful, despite their ridiculous costumes, and I felt better too.

Next to the market a coconut vendor was setting up shop, pulling a tarpaulin off a mound of what looked like green volleyballs, laying out a styrofoam chest of soft drinks and a machete for hacking the coconuts open. He was singing; he must have felt good too. Fishing boats were right side up again and being slid into the water by teams of men, two racing from stern to stem to replace the wooden rollers. In the distance, the reef was a white line drawn on the turquoise sea. Things were looking up in Maceió. The hotel's coffee came thick and black from an ancient three-legged urn. When Fátima rolled up I was already outside with camera and notebook.

"It's a long day," Fátima said. "I remembered you like music. So I brought some." I couldn't remember discussing music or anything else with Fátima. There was a container of CDs the size of a lunch hamper between us.

"Yes, you have," I said.

"*Lots* of music," said Fátima. She backed suddenly into the stream of traffic, which stopped miraculously to let her in. Then we shot forward along the causeway. There was an odd smell in the car, something I couldn't place but that inexplicably took me back to my days as a biology student. I cranked the window down.

"You like this beach?" Fátima waved out her window without looking. She didn't seem to notice the smell.

"I love beaches," I said.

Fishing boat near Maceió, Alagoas.

"It's a terrible beach. Later, if we have time, I will take you some-where better, much better." The car filled with manic-sounding music, a flute and a clarinet twirling together above many guitars. In Brazil, one guitar is good, but four guitars is better. "You like it? We can buy it later." Fátima slapped the steering wheel and we flew out of town, trailing notes into the warm wind. On my left, the sight-seeing outriggers were already heading out to the reef. They slid sideways as they crawled toward the thin line of surf, making almost as much leeway as forward progress. The morning sun lit up the yellow Skol Beer logos on their triangular sails.

On the outskirts we stopped at Fátima's laboratory, in a former military academy taken over by the university. While Fátima col-lected buckets, I estimated some distances in the crumbling old complex: six-metre-high doors, dark wooden flooring where each board was forty centimetres wide. The windows all had iron grat-ings and the upper floors had balconies. If you had to spend your time bent over a microscope counting the annual rings in fish scales, this was the place to lend a certain *gravitas* to your work. In the gardens, ornamental trees were surrounded by little rings of plastic pop bottles half-buried in the ground.

South of the city, the ocean road narrowed to two lanes. You could almost imagine the place five hundred years ago. Way out to sea, that dot of a fishing boat might have been a Portuguese slaver picking her way along the reef, looking for the way in. From the ocean, it must surely have looked the same then; the crew would have seen the same green expanse of sugarcane we were driving past now. The fields ran uninterrupted on either side of the car, the ocean only visible as fleeting glimpses down service roads. At harvest time, Fátima told me, you could earn twenty-four reais a day — about $12 — cutting cane. It was the worst kind of work: hot and smoky from the fires razing the spent fields, machetes swinging in fingers slippery with sweat, snakes slithering away as you advanced. Land changes hands slowly here, and the sugar business is still controlled by just a few families, as it always has been.

A few people stood by the side of the highway holding out plastic bags of beans, and the shadows of circling vultures swept across the pavement; when you looked up and spotted one overhead, the sun sparkled through the trailing feathers on its one-metre wings. The vultures seemed to be doing well. At one spot there were twenty of them in an ugly cabal by the side of the road, tearing at the green body of a metre-long snake. As we drove, Fátima entertained me with the latest scandal from upriver. Back closer to the Sobradinho dam, an agricultural enterprise had obtained funding through the Bank of Brazil to grow manioc, an unlikely choice for making money but nobody had looked too closely and the request had been granted. A year later the manioc was identified as marijuana (the plants do look kind of similar). We wouldn't be going there, she said, shaking her finger at me. Fátima was already turning out to be a good guide, much more entertaining than the President of Alagoas.

We reached the town of Piaçabuçu in about an hour — the place where Frei Cappio's long march had finally ended, where he and his companions had poured out the water they had carried from the river's source. The São Francisco enters the sea here, and for such an important river the entry was unimpressive, as though an understudy had taken the stage at the last minute. It was October, and the lack of rain combined with the strangulation of the dams had turned what should have been a headlong rush into an exhausted stumble. The estuary, Fátima told me, was becoming saltier as river flow declined, and it was true, the only boats I could see were shrimpers. Now, a "salt wedge" can get as high as ten kilometres up the São Francisco. A fish called *robalo,* a sea bass, had been found fifty kilometres upstream, confused and coughing, its gills working overtime.

In the old days, the carbon signature of the São Francisco extended fifty kilometres from the mouth. The plumes of great rivers are visible from space, and the São Francisco still has a plume — but it's the wrong one. Now, silt that used to be brought down from high up in the *sertão* is scoured instead from the river mouth itself because, by the time the São Francisco has passed through the chain of dams, whatever solids it's been carrying have sunk in the calm waters of the reservoirs. So: the reservoirs fill up and the river turns clear and barren. And where the river meets the sea — what the geologists and hydrographers call the "marine end-member" — the mixing zone has lost all semblance of the natural pattern.

The town of Piaçabuçu seemed to be surviving on tourism. I picked up some pamphlets from the ex-secretary of the environment who was now running an ecotourism outfit, and we headed the twenty-five kilometres upstream to the larger city of Penedo, the last point on the river you can visit easily by car. After Penedo, the road leaves the river and you have to zigzag, connecting the dots of tiny towns and stitching your way upriver until you reach the first big dam at Paulo Afonso.

Penedo itself is built high on a bluff, so that from the centre square you look down narrow streets through Portuguese colonial façades to see the river. Six-metre cacti dot the lower promenade like spiky candelabras, reminding you how close you are to the arid interior. Penedo is one of the architectural treasures of Brazil, but it's small and off the beaten track and it doesn't get the attention of the gold rush towns like Diamantina and Ouro Preto. But there are the same cobblestone streets that moan beneath your tires, and the place was obviously once a centre of power, with a commanding view of the river. An ancient cannon still guards the town, aimed downhill at the water and rusted in place. There weren't many boats down there. We did a U-turn around the square and headed inland, back-tracking to the place Fátima really wanted to show me — Marituba. The body of the patient was dying, and Marituba was its lungs.

Marituba is a *várzea,* a floodplain, straddling the last tributary of the São Francisco just before it enters the Atlantic. In a big, tropical river like the São Francisco, the *várzea* is the part everybody seems to forget about. If they didn't, they'd have to adjust their thinking to accept the fact that their river isn't just a channel of water between banks a kilometre apart, but an extremely leaky pathway in and out of which the water meanders depending on the time of year. When it rains, channels open and the *várzea* fills like a vast and breathing sponge, a green maze of connecting lagoons that are the key to sur-vival of most of the fish species in the river.

The migratory fish like the *surubim* from the telephone booth in Maceió need the *várzeas* to survive. They enter its interconnecting channels on the floods that follow the rains and fill them with new-spawned eggs. Then the *várzeas* become vast nurseries. The water is warm, and embryonic development of the newly laid eggs is explo-sive: cells divide furiously, eyes appear, tiny transparent hearts start to contract — all in twenty-four hours. The newly emerged fry feed voraciously in the highly productive waters. On the Velhas River, Burton says, "The pilots speak of sixteen to twenty palms rise, and of small bayous, more often flood-lagoons than filtration lagoons, formed in the flats." Those were the *várzeas.* They are the world's most threatened ecosystem.

Humans were never meant to live on unstable rivers like these. Flooding is a big part of that instability; Burton devotes pages and pages to it. Everywhere he notes the watermarks on houses: three feet, six feet, ten. He meets a property owner and sees a tree marked by the last big flood, forty feet above the present level: "People were taken by canoes off their thatched roofs." At Guiacui, Burton describes clay banks "twenty-nine feet six inches high" (I love that "six inches") with the walls of the houses showing a watermark more than six feet up. That's a ten-metre flood. In the town of Januária, he wrote about a church, Nossa Senhora das Dores: "In

1843 there was an inundation in which a *surubim* was caught in the church." A catfish caught in a church! What sticks with me is not the fact of the brute swimming in through the window, but that somebody actually *caught* it. *Carpe diem*, I guess.

Eventually, people learn to control floods. Regulating such a river for hydropower and flood control, and taking water out for irrigation, mean that the flooding of the *várzea* is up to man, not nature. The temptation to plant grapes or eucalyptus, or to invite cattle into all that conveniently located land, is overwhelming. And so the lungs become clogged or cut off from the body and the river turns into a channel, nothing more. This is what has happened to most of the São Francisco.

The Marituba *várzea* is small, but at least it has protection of sorts as an Environmental Preservation Area. Fátima's real work was here, collecting specimens of the local fish to see what was thriving and what was gone, and to test them for contamination with heavy metals. When we arrived in Marituba, she was obviously in her element, taking the Fiat down a maze of red dirt roads, pulling up in a cloud of dust so that I could watch a man poling a bright yellow dugout canoe through what looked like a highway of lilies. There was water under him, lots of it, but I could see no more than the occasional sparkle from the sun. He might as well have been on land. In his jaunty white cap he looked like a Venetian boatman. Farther on, Fátima stopped again, this time in front of a cluster of small houses, and before we were even out of the car a woman had emerged to greet us.

Didi was black and sturdy-looking, as though it would take a lot to knock her off her pins. She might have been forty; she might have been seventy. She wore a white T-shirt, a bilious green and yellow "Brasil #1" baseball cap and a faded and flowered red dress that fell below her knees. In one hand was a sealed plastic bucket, the kind institutions buy ice cream in. Didi beamed and clapped me on the back while Fátima did the introductions. She seemed to be apologizing to Fátima for something, but a strong accent and many missing teeth made her hard to follow.

"Her husband's drunk." Fátima translated Didi's dialect into her own citified Portuguese. "But we can meet her brother." Out came

Demonstrating a fish trap in the Marituba wetland.

a wiry man with steel-grey hair and the body of a twenty-five-year-old. He pried the lid off the bucket. Fátima bent over and rummaged through the liquid to extract an eel-shaped creature a foot long, curled unnaturally like a ram's horn. *"Synbranchus marmoratus!"* she cried, like some kind of celestial invocation. She flung her arms around Didi, the pickled eel dripping behind Didi's back. The faint smell that had been bothering me all the way from Maceió was much stronger now: formaldehyde. Just like that, I was ten years old again, back in the bottle room of the British Columbia Museum. Fátima pushed the thing back into the bucket and went to the car, opened the trunk and extracted an identical pail, swapping it for the one with the eel.

"Didi collects fish for me," she said, wiping her hands on her jeans.

We followed sister and brother to the lagoon, Didi swinging a fish trap from one hand. It was a cone-shaped basket, open-ended, a design used for centuries. Simplicity itself, she demonstrated, wading into the shallows and dropping the trap at her feet. She mimed reaching in and pulling out a fish, but I was already looking past her, to where the lagoon opened out into a broad patch of clear water. I could see some floating structures, four of them at least, like waterlogged swimming rafts.

"Tell me that isn't what I think it is," I said to Fátima. "Not here." It wasn't a raft, I could see that now; the part you could walk on was only a narrow square around the perimeter. The nets would be inside.

"It is," said Fátima. "The government tells the people it's better to farm the fish, so they give them the equipment, the baby fish, the training." She pointed to a dugout nosed into the reeds and shrugged. "You just paddle out there and feed them."

"Them being what, tilapia?"

"*Claro,* tilapia. Our future."

Marituba is becoming managed. One of the main tools is the same one unwittingly embraced by Miss Tojo and her friends when they sit down to dinner: stocking water bodies with hatchery-bred fingerlings. Tilapia farming is a form of stocking, because of course the fish escape the nets and breed in the wild, but stocking the *várzea* with native species is just as damaging genetically.

In Brazil, irony and fatalism go hand in hand. You need one to deal with the other. Even as the *várzea* is being altered by introducing alien species, new species are being discovered. Richard Burton was right when he said, "The naturalist who shall attempt the ichthyology of the São Francisco will have before him a task of years." (Or, in Fátima's case, "have before her.") It takes a special kind of person to come to a place like Marituba and do what Fátima does, shrugging off the evidence that larger forces are against you. As irony goes, the scene here was hard to beat. Marituba was an officially protected area, fragile, and home to numberless endemic species like the eel curled up in Fátima's bucket, but her research clearly showed that the dominant species in the *várzea* was now the disc-shaped *Metynnis mola,* the "CD fish" — a native of the Amazon. Nobody knows how it got there.

Now, people who had coexisted with the native species were acquiescing in the government's official counter-offer to make up for their loss: farming an introduced fish notorious for its ability to overrun aquatic ecosystems and push out the locals. Tilapia grow fast and cheaply. They taste great. In Colombia, people slash them, dredge them in salt and deep-fry: with a wedge of lime, delicious. To government agencies committed simply to replacing lost protein, they are the no-brainer solution. I'd seen cages like this in the reservoir near Três Marias, two thousand kilometres upriver, and I knew they were in Sobradinho too, rickety structures ready to spill their cargoes into the open waters at the first big storm. I paddled out to a raft of them near Três Marias once; when I climbed onto the narrow deck the whole thing groaned and my feet were suddenly awash in warm river water. Farmed fish escape; it's a given. The fact that this African fish is an invasive, alien species isn't stopping anyone from loading them into flimsy cages in Marituba.

Government agencies — especially departments of agriculture — have invested heavily in tilapia culture, creating better breeds, better diets, better infrastructure. It's a development dream, a fish that in some parts of Asia is now cheaper than chicken but tasty enough to be exported, given a fancier name and sold in a market in London for ten times that amount. Tilapia even clicked for a while in Japan, before Miss Tojo and her friends got tired of it. In Venezuela ten years earlier I'd found tilapia in a lagoon in the Orinoco — "the university fish," the locals laughingly called them. They probably aren't laughing now; *pesce de Universidad* might be all they have to eat.

Tilapia are phenomenal survivors and it's hard to argue with the ease of sticking them in a net and watching them grow. They can even live in seawater, so a *várzea* where reduced freshwater flows encourage higher salinity is no deterrent. Joggers along the Ala Moana canal in Waikiki can look down on carpets of mud pockmarked with tilapia nests, like shell craters in a battlefield. How did they get there? Escapes from freshwater ponds. The only thing really limiting the spread of tilapia is their temperature tolerance, because they're still a tropical fish — so far. But it won't be long before some researcher sticks in a gene that allows them to thrive next door to trout. A nice bit of science, but an ecological catastrophe.

SIMON AND GARFUNKEL

We said goodbye to Didi and her brother. Mr. Didi continued to sleep it off somewhere out of sight. Fátima settled the bucket of new specimens in the trunk, where it would no doubt leak like its predecessor, and we headed out of the *várzea*. It was twilight, and by the time we were back on the coast road to Maceió the sky and the land had already begun to reverse roles, the hills silhouetted against the lighter grey of the sky. Fátima pulled over.

"I don't see so good at night," she said. She pointed to the glasses hanging from a cord around her neck. "Middle age, *né?* You drive."

We circled around opposite ends of the car and she fiddled with the CD player while I fumbled with the seat. There didn't seem to be anyone else on the road. I found the switch for the headlights and pulled out as a Chopin *Nocturne* filled the car. Something mournful and aptly named; Fátima knew her musical stuff. My thoughts went back to the river, to *várzeas* plugged and abandoned, then repopulated with an off-the-shelf African species that was collectively a lot smarter than the well-meaning bureaucrats who put it there. We hit a bump and the bucket in the back shifted. I thought about people like Fátima, herself a drop in another kind of bucket, doggedly collecting specimens that would end up pickled and stinking and curled up, all that was left of a phenomenal diversity of species. There wouldn't be anywhere to store her collection safely; sooner or later the buckets would crack and start to weep formaldehyde and someone following the smell would pitch them out. The same fate as my beloved bottle room. I glanced over at Fátima. She was turned away from me, staring up at the night sky.

And now something was wrong with the headlights. There seemed to be only one, and it pointed left, into the oncoming lane. I scrabbled with the switches on the steering column and finally found the dimmer in the old-fashioned place, next to my left foot. I pushed it — and the single light went out. We sailed along for a few seconds in complete darkness before I got it back on again.

"Um, Fátima," I began, but Fátima was fiddling with the CD player again. She had a lot of music and a short attention span. Three apparitions suddenly materialized on the right: a person, a bicycle, a dog. "Jesus," I said. A road sign loomed in and out of view: Maceió, 55 km. I forgot about *várzeas*. In another ten minutes it was pitch-black, the tropical night enveloping us with the suddenness that always takes a northerner by surprise. I snuck a look at my watch; seven o'clock, but it might as well have been midnight. More automotive problems began to emerge, like night creatures. The car was horribly underpowered. When I downshifted to climb one of the gentle coastal hills I had hardly noticed on the way out, the engine screamed. I imagined smoking fan belts and melting pistons.

Headlights began to appear in my rearview mirror, expanding with horrifying speed until the car filled with light and for a few moments I could actually see where I was going. When each car passed I drove by its tail lights until they too vanished, leaving us alone again in a tunnel between walls of sugar cane. Without the reference points of the other cars I floundered repeatedly onto the shoulder. There were no markings on the road. It was another "driving in Brazil" story. Maybe worse than the animal hazards of the Pantanal — caiman, capivara, anteaters — maybe better than the potholes near Pirapora and their way of lying in wait on a dappled road. But no fun.

Now even the soundtrack was falling apart. Fátima's interest in music was all-encompassing but she listened like a bee in a field of clover, stabbing at the invisible console, extracting discs, shoving new ones in their place. Chopin got cut off in mid-arpeggio and replaced by Simon and Garfunkel. I gritted my teeth and a truck bore down on us from the other direction, its lights flashing irritatedly; another car appeared suddenly on our tail; a bicycle wobbled by on the shoulder — what were people doing out there on bicycles at this time of night? Simon (or maybe it was Garfunkel) whined about darkness. Of all the ways to die, I thought. Please, don't let them find my broken body with this music still playing.

Finally the bicyclists and dogs and families thickened to a steady stream, lines appeared on the road, the lights of the city washed out the sky ahead and Fátima awoke from her convenient reverie to issue directions. She had chosen a restaurant. It was somewhere in the middle of the city and I am sure I could never find it again, although I remember that parking was near-impossible. Parking in big Brazilian cities has some subtle rules, especially in the evening, near restaurants. There's usually a guy hanging around the street; you give him a few reais and he "watches" your car. If you don't, maybe he forgets to watch it. I like the system, because it's cheaper than paying a parking company and it's quintessentially human, a fine balance of trust and apprehension. We surrendered our exhausted car to a kid in a dirt lot that was rapidly filling with BMWs and Benzes. Prosperous-looking young people headed across the street.

"Gula Divina," said Fátima, pointing to the sign above the entrance and tugging at my elbow. I didn't know *gula*. "It means . . ." she searched for the English word, "divine overeating, I think."

"We can go in there?" I said. "Like this?" I was wearing a stained T-shirt with a design of sperm cells; someone had given it to me at an agricultural conference years ago. My runners were caked with Marituba mud.

"Why not?" said Fátima. I began to wonder if she was a supreme ironist or whether there were Brazilian realities I had absolutely no inkling of; both, probably. I knew she hadn't washed her hands because she still trailed a faint odour of formaldehyde. We picked our way through din and smoke, dodging waiters bearing flaming grills at shoulder height. The air was filled with the sizzle and pop of cooking meat. We breathed an aerosol of fats and juices and expensive body lotions. Across from our table two mountainous young women laughed and drank and forked *picanha* — the tender-loin or "queen of meats" — into their mouths. They looked inflated, their brown shoulders smooth as butter.

Ten minutes away, I knew, there was a *favela,* Pontal de Barra. There, the brown bellies of the young women were swollen with children, not steak; they lived in shacks of corrugated metal and cardboard next to a filthy canal, beneath the white silos of the chlorine factory that made their eyes sting. The men fought or begged or fished from canoes in a lagoon where the water was a lurid algal green, the hydrocarbons on its surface catching the sun in iridescent swirls. And beyond the *favela,* a couple of hours down the coast road, there was Marituba, where people still fished with baskets and poled their canoes through wetlands infiltrated by African tilapia, looking for species living on borrowed time. More infuriating ironies. We settled in and ate like loggers.

SHEER FOLLY

Back in my hotel room, I carried a plastic chair out onto the deck. The strip was quiet except for a lone sound-truck making the rounds, a Volkswagen van encrusted with speakers the size of bookshelves. Wagner would appear on the beach tomorrow. Wagner the Brazilian pop star, I assumed, not the composer of "The Ride of the Valkyries." But it was hard not to feel apocalyptic after a day like today: too many worlds too far apart and yet too close; too many contrasts; too many reminders of the preciousness and cheapness of life.

Any development worker knows that many poor countries aren't poor in any absolute sense; instead, they just fail miserably at distributing wealth. Brazil is thus fabulously rich and miserably poor at the same time: *Gula Divina* and pregnant fifteen-year-olds in the *favela*. Hardly a novel observation, but it can be debilitating if your job is somehow to "help," because the forest rapidly gets lost in favour of so many trees calling for attention. "Think globally, act locally" is a wonderful phrase, and I believe in it, but there are times when you can't just flick a switch and change your frame of reference. What's the right response when you have to get up the next morning and do your job? Cynicism? World-weariness? Tears?

I rummaged in my pack for the book I had brought for just such an occasion. *The March of Folly* was written in 1984 by the historian Barbara Tuchman. I read it back then and, ever since, have kept in some corner of my mind the memory of Tuchman's exasperation at the perversity of people following paths that guaranteed their own annihilation. The voice of reason, I thought now, just the ticket and not only for what I'd seen today but for all the insanity on the São Francisco River. I took the book across the street to the kiosk and sat down at a rickety metal table with a cold beer. The sound-truck inched past, howling about Wagner.

"Mankind," Tuchman began, "seems to make a poorer performance of government than of almost any other human activity." And,

"Why do holders of high office so often act contrary to the way reason points and enlightened self-interest suggests?"

I felt better already. Tuchman was going to sort it out for me. Why, she wondered, would the Trojan rulers drag a suspicious-looking wooden horse inside their walls? Why did Montezuma roll over in the face of a few hundred raggedy Spaniards who were plainly human beings, not the gods he feared? Why, I added to myself, would the government of Brazil insist on pursuing the lunatic scheme I kept hearing about, to tap the São Francisco and divert precious water through thousands of miles of concrete channels in the name of poverty alleviation — when there were easier and cheaper solutions?

Pictures taken of Tuchman around the time *The March of Folly* was published show a laughing woman in her prime, with strong, almost mannish features and wavy silver hair thrown back. She reminds me, weirdly, of the former Japanese Prime Minister Koizumi, and although her skewering of fools is relentless she's not the finger-wagging kind. "Oh, *hell*," I can hear her saying when the shenanigans of the popes or the American administration become too much for her, pushing the Smith Corona aside and striding down to the dock in front of her New England cottage. I imagine her with tiller in one hand and mainsheet in the other, tacking across Chesapeake Bay to clear her head.

Tuchman's prose was lucid, her arguments unimpeachable. Governments do plain nutty things, she was saying to me now, and she laid out her criteria for folly: the action had to be seen as counter-productive even in its own time, it had to persist beyond a single political lifetime and in the face of urgent warnings, and a feasible alternative had to exist. The São Francisco diversion fit these standards perfectly. Wooden-headedness, the refusal to learn from experience, it had all the symptoms. I finished my beer, watched by a rail-thin woman dragging a plastic sack. When I got up to return to the hotel she darted forward and snatched the can off the table.

I got into bed and read on. When I finally fell asleep I was thinking not about rivers and pollution and overfed Brazilians but about Montezuma and his court and how they had confronted the appearance of the Spanish adventurer Hernan Cortes in 1519, at almost exactly the same time as the Portuguese had landed not so far from where I was now. Cortes had three hundred men and Montezuma was the ruler of the Aztec Empire's five million, but it was the Aztec who lost everything. Montezuma's folly, in which he was supported by enough of his counsellors to make defeat inevitable, was to ignore the all-too-obvious evidence of the invaders' mortality — even the severed head of one of their soldiers. Instead, he preferred to believe that Cortes was the fallen god Quetzalcoatl, returned finally to portend the downfall of the Aztec empire, despite the obvious evidence that the visitors worshipped, as Tuchman puts it, "a naked man pinned to crossed sticks of wood." By the time Montezuma was discredited among his own people, the Spaniards had made the alliances they needed to control what is now Mexico City, rebuilding it and stamping Spanish rule upon the country for the next three hundred years. Cortes and his men should have been squashed like ants; thanks to the folly of Montezuma and his court and their wooden-headed belief in the return of a vengeful god, Spain took the country in a cakewalk.

Tuchman's retelling of the fall of Mexico is only her preface, her warm-up to even more spectacular follies to come, but already her elegant prose was an antidote to the sense of things flying apart that my visit to Marituba had engendered. Today had ranged from the sublime to the ridiculous: the gasping river from which even more water was to be extracted and sent to the arid northeast, a place I had never seen but that had already taken up residence in a corner of my mind; the tilapia cages; our blundering back in the cyclopean Fiat; the bacchanalia in *Gula Divina*. But her story substituted other images of chaos, and I went to sleep imagining smoke and sizzle of another kind, and screams, as the army of Cortes hacked their way into the last capital of the Aztecs.

CHAPTER 12

THE CASTIGATION
OF GOD

*Drought in northeast Brazil, and how to get rich
from it*

Before the rains: an açude, *or small reservoir, in Paraíba, northeast
Brazil.*

THE HISTORY OF NORTHEASTERN BRAZIL ISN'T PRETTY, OR SIMPLE. There are so many elements, although some, at least, are familiar. The loss of indigenous lands and customs, for example. I grew up in another place where European colonists ran roughshod over the local Indians, crowding them into reservations, taking their land, outlawing their customs. But there was no official genocide, and the land that was stolen was Eden, not desert. The North American Indians were cheated and infected and systematically discriminated against but they weren't wiped out, and today they have at least a firm legal and moral toehold on getting back some of what should rightfully be theirs. Resettlement, too: for the most part, individual North America settlers got title to their lands, up front. Not so in Brazil. Next, slavery: other countries have had this despicable institution, and for the same economic reasons as colonial Brazil. But what happened ethnically in Brazil was more complicated than in, say, the United States, where interbreeding and the mixing of cultures were less common. Brazil is hardly the paragon of racial harmony it's sometimes made out to be (racism just exists in more subtle forms, the commonest being advancement based on your actual shade of blackness unless you're an artist or an athlete), but it has undeniably been a melting pot.

Feudalism is another parallel: it worked well in the new colony of Brazil, and it hasn't entirely disappeared. Most of the landless people in the region are squatters or sharecroppers. Drought simply aggravates the problem, like a thumb in a wound. The valley of the São

Francisco, especially its arid middle and lower sections, has been Brazil's crucible for endless battles over water and the constant widening of the gap between those who have it and those who don't.

But to really understand the northeast, parallels with one's own culture need to be jettisoned. You have to go back to first principles — to geography. Look at a map of South America. Brazil takes up a good two thirds of it. The lump that sticks like a sloping shoulder into the Atlantic is the northeast. No other section of the country gets its own geographic moniker; you don't hear anyone speaking about "the southeast" or "the northwest." But in development circles, "northeast Brazil" is code for the largest pocket of misery in the Americas. Richard Burton penetrated the northeast through Bahia after running the São Francisco from the comparatively verdant hill country of Minas Gerais, and he called it "this sandy thirsty land, rich only in thorns." But he also said, "It wants only water to become as fertile as Sindh," a sentiment that reverberates the length of the São Francisco to this day.

What's special about the northeast? If your map is one of those useful ones with different colours for different ecosystems, the kind you get as inserts in the *National Geographic* and can never find a good place to store and eventually lose just when you finally need them, you can get a clue just from the colours. Plenty of Brazil is green, and the Amazon region is the deepest, most vibrant green of all. But the northeast, once you leave behind the green fringe around the coast and the states of Maranhão and Piauí that border Amazonia, is an exhausted brown. Twenty-five million people live there, about one eighth of the country's population, and it's *dry*. Technically, the climate is "semi-arid," which means it's enragingly variable. The fatalism at the core of Brazilian culture derives from the northeast, where annual rainy seasons can be suddenly and arbitrarily suspended. From July to September it's *always* dry; every decade or so, El Niño visits upon the area a drought so severe that some areas won't see the regular rains for three years. Based on the number of people, the size of the area and the amount of rainfall it gets, the Brazilian northeast is the most densely populated semi-arid region on the planet.

Brazilians call the driest part of the northeast the *polygona de seca,* the "polygon of drought." It's one of those wonderful names I've always found tinged with mystery and dread, like the Mountains of the Moon or the Bermuda Triangle. But if you ink in the nine states in the so-called polygon you have to strain to see any straight sides, so it's not a polygon at all, just a big blob on the shoulder of Brazil. Most of it sits on crystalline, impervious bedrock, and the evaporation rate is more than double the amount of rain that falls. When rainwater comes, the thin soil is poor at soaking it up, so it runs off or evaporates, a process that happens fiercely and fast. When I first visited the northeast I hung a wet cotton T-shirt out the window of my hotel, and it was dry in an hour.

Scientists use a wonderful term, "extreme hydrological events," for what happens here: droughts, floods; the whole place is like a person with borderline personality disorder. Agricultural soil is shallow. Groundwater exists, enough at least for the needs of small farmers and communities, but it's sometimes hard to find. The few larger deposits, like those tapped by the large coastal cities of Recife and Natal, or others used heavily for irrigation throughout the entire São Francisco Valley, are already over-exploited. Many of the smaller aquifers in the northeast are brackish and need treatment before they can be used for drinking.

For usable water, that leaves the rivers. But with one big exception, most of the tributaries in the region are intermittent, drying up for months or years. The big exception is the São Francisco, which starts way back in Minas Gerais, where there's plenty of water, and runs right through the northeast all the way to the Atlantic. Sixty per cent of Velho Chico penetrates the polygon of drought, an artery in a moribund limb. Drought is worst in the lower forearm and wrist, in the states of Bahia and Pernambuco and extending from the immense Sobradinho reservoir (which didn't exist in Richard Burton's time) to the falls at Paulo Afonso (which did, although they were so big then that he called them "Brazil's Niagara"). Vegetation is scrub savannah, mostly cactus and small bushes called *caatinga* — "white bush" in the Tupi language — and the most important economic sectors are irrigated agriculture and hydroelectric power. Until recently, most of the large-scale agriculture and hydroelectric development on the river took place in this

A carpet of cactus, Paraíba, Brazil.

section, an arena for water conflict ever since the Portuguese established the capital of their new colony in Salvador da Bahia in 1549.

The São Francisco has been called the "Brazilian Nile," after another long river that flows through a drought-stricken region. Brazilians themselves refer to it as the River of National Unity, a name that's usually explained as representing the river's role in taming the semi-arid backlands over the last five centuries. Unity, though, seems to me the last thing that's happening around the river now. River of social exclusion might be more like it. To understand that exclusion — where it came from, why it persists — it helps to know the story of Canudos.

SLAVE CITY

The War of the End of the World is the title of a novel by the Peruvian writer Mario Vargas Llosa. Based on real events and people, it's the story of the doomed "rebel" city of Canudos, in the backlands of Bahia, a story of landlessness, fatalism and repression, the dark currents that course through the lower third of the São Francisco basin in northeast Brazil. To know something of Canudos is to understand why, just over a hundred years after the last house was razed and the last throat slit by the army of the Republic, resistance to the megaproject that will divert the river is so passionate.

By 1877, Richard Burton had been gone from Brazil for ten years. He was in Trieste now, at his final posting, his mind on his translation of the *Thousand Nights*, not the backlands of northeast Brazil. But the *sertão* was still there, and in 1877 it didn't rain at all, nor the year after. Three hundred thousand people died. Homes were abandoned. The drought brought epidemics; women miscarried, children's hair and teeth fell out, adults swelled with tumours and spat blood. Rattlesnakes appeared by the thousands so that people had to arm themselves with clubs and machetes. Vargas Llosa's novel is a flood of images of the life of the *nordestino* in this horrible time: men in sandals, leather hats and drill trousers, *cantadores* wandering over the countryside telling tales of suffering and abandonment. Gangs of outlaws multiplied "like the biblical loaves and fishes."

Mass migrations followed. Many of those on the move were looking for land. After 1870, land titles that were unregistered were declared void by government, beginning the informal "occupation" of land by *sertanejos* (inhabitants of the backlands) trying to escape a life of servitude to the big ranchers. They set up a communal lifestyle in these settlements that persists today in many parts of the northeast and has become the legal basis for official land reform. The climate of those times was ripe for a visionary, and one appeared: Antônio Conselheiro. He had already wandered the backlands of Sergipe and Bahia for a decade, preaching to the poor; he was a dangerous mixture of mystic, saint, fanatic. People began to follow the man they called The Counsellor on his pilgrimage through the *sertão*, drawn by his conviction that the Republican government that had displaced the monarchy would be worse than the drought, its tax collectors greedier than the vultures and bandits. The number of pilgrims swelled after the abolition of slavery in 1888, and the fatalism of generations castigated by drought and disease found a home in what the new government called religious fanaticism one day, sedition the next.

Conselheiro became a marked man. In 1893, he and his band settled in Canudos, a town in northern Bahia not far from the São Francisco. Canudos represented hope, health, refuge, forgiveness; there, a collection of landless peasants, halfbreeds and reformed bandits established a settlement that was to last four years. Most were freed slaves who had abandoned the plantations. Their communal lifestyle infuriated ranchers, who feared losing their workers, and it worried the Republican government, who saw Canudos as a monarchist challenge. It's not clear how much of the revolt was political and how much religious. The Counsellor preached Christian values within a strange system where the Antichrist was the Republic, and the only currency permitted within Canudos was coins bearing the likeness of the ex-Emperor Dom Pedro.

The followers of The Counsellor fought hard for their vision, many of them with the savagery learned in their former lives. One such was Big João, the offspring of slave parents bred by their master on a sugarcane plantation. João was taken in by the mistress of the plantation and treated as a pet. One day he turned on his mistress in her carriage, assaulting her unspeakably before slaughtering and dismembering her. "I've got the Dog in me," he said. A life of banditry followed, during which he earned the nickname of "Satan João," until the fateful moment when he met The Counsellor.

But even a force led by a man as pitiless as Satan João could not hold out forever. Twelve thousand soldiers from seventeen states — half the Brazilian army — attacked Canudos in four expeditions, each one more complex and encircling than the last. In the final assault, the army brought everything it had, including its new Krupp cannons and machine guns. Seizing the water supply was decisive, a tactic that drove the women and children to suicidal nighttime sorties to the heavily guarded wells. Vargas Llosa dramatizes the sacrifice they made for that most precious of things, the soldiers "posting themselves at Fazenda Velha and waiting for the light of the moon and the shadows creeping up to get water . . . in the morning the ground around the wells was strewn with bodies."

Resistance guttered out on October 4, 1897, the day of St. Francis, after five days of siege. Twenty-five thousand people were killed in battle. Most of those taken prisoner were murdered, the favoured method of execution being the *gravata vermelha* or "red tie" — a slashed throat.

Canudos was razed and burned. It had been, briefly, the second largest city in Bahia. The President of the Republic declared, "In Canudos, nothing is left standing now." A second Canudos arose a decade later, but that one was obliterated too, when a dam was built in 1969. The Rio Vaza-Barris that had fed the settlers became the Cocorobo Reservoir, devouring houses, ruins, and memories in a mixing of water and blood.

The defeat of Antônio Conselheiro sent a powerful message to landless people tempted to challenge the established order in Brazil. Dependency, the government was saying, would not be easily dislodged. The Republic might be new, but it was quite happy to coexist with the rural oligarchies established by the first colonists. Business would go on. Flooding the ruins of Canudos under the Cocorobo reservoir just rubbed it in. But Brazilians, especially *nordestinos,* have never forgotten Canudos. Conselheiro's movement was put down with such savagery it has remained forever the symbol of the crushing power of the state. There have been twenty-nine films, most of them documentaries, more than two hundred books, as well as songs and poetry written on the subject. Popular celebrations for the martyrs of Canudos are held every year. The writer Ana Augusta Rocha wrote of "men who searched for rain and instead became flooded with sun." For her it was a place of "burnt retinas and blind faith."

THE IMPOSSIBILITY OF OWNING LAND

The northeast sticks out into the Atlantic like an invitation, or at least an obstacle big enough that colonists from Europe had inevitably to stumble ashore. Settlement of Brazil followed a geographic pattern that reflected booms and busts in crops. The Portuguese started in the coastal northeast with sugar. The big cities

— Salvador (1549), Recife (1537), Fortaleza (1611) — are all on the coast. Cane persists there, along with tropical fruits and palms, but once you get inland it's only the tougher plants that survive. Some are industrial, like cotton, agave and sisal; others are grown mainly for subsistence, like corn, beans and squash. The job of working the land, first to grow sugar cane and then later to raise cattle, was done by slaves.

The first slaves were Indians from the interior, hunted down on horseback by the *bandeirantes* (flag-bearers) who are a savage chapter in Brazil's history. In their heyday the *bandeirantes* killed, laid waste, prospected and interbred, and the slaves they brought back sufficed to kick-start the new colony until a cheaper and hardier alternative arrived: Africans that could be had in bulk from across the Atlantic. Miscegenation was the official policy of the colonizer for hundreds of years, the Portuguese mixing first with their Indian slaves, then with their black ones. The result was a landless, mixed-race class who worked the holdings that belonged to the fortunate few. This is how Richard Burton describes the cattle-herders of Minas and Bahia: "Wild as the Somal . . . sitting loosely upon ragged nags . . . huge spurs around their naked heels."

The São Francisco was a main route for penetrating the interior, and once the *bandeirantes'* time had passed, prospectors and miners used it to get to the gold of Minas Gerais. Later, in the nineteenth century, the river became a transportation route in the other direction, carrying goods from the interior to the coast. When the sugar-growing areas were depleted and competition from the West Indies became too much, coffee followed, on the treed slopes of São Paulo and Minas Gerais. Coffee brought a new wave of immigrants, mostly Italians, to replace the slaves finally freed in 1828.

Like most new countries established by an overseas colonial power, much of Brazil got divvied up based on who was first in line and knew the best people. Connected families from the old country found themselves with vast land grants or *capitânias* (captaincies), many of which persist to this day as immense *latifundias* or tracts owned by people who live far away, in the large cities. When the first colonists came to the *sertão* they found only a place to graze cattle, but they took advantage of the space and their ranches were immense. At the beginning of the eighteenth century, for example, the Garcio D'Avila family owned three hundred and forty square leagues along the São Francisco — about seventy-five hundred square kilometres. Many *latifundias* owe their existence to illegal registration where public land is transferred to private holdings. Brazilians call it *grilagem* — putting a cricket in a box filled with faked documents. The cricket eats and shits and moves around, the documents look older and older until *voilà*, land title.

Concessions from the Crown, and then from subsequent imperial regimes, firmly implanted the belief that *land was a right* — but only for a few. Working that land was the role of Indians, blacks, poor Portuguese immigrants, and all the colours of the rainbow that they produced. Slavery ended officially in 1888 but the feudal system persisted, and so did the mentality of subjection and dependence. Official policies of exclusion, slavery and genocide combined to create a vast and landless underclass of whites, Indians and escaped slaves. Occasionally, all these elements came together in popular movements (Canudos was the most famous); always, there was poverty, suffering and violence. Most people lived on the land, but they did so by the good graces of the owner, and they paid for the privilege in food. The Land Law of 1850 simply formalized the original royal land grants, and most of the interior continued to be given over to cattle farming. People worked the herds for the owner or survived by sharecropping, but they never legally owned their land.

Sertanejos — those who live in the backlands — have always been survivors. There is history in their faces, the raw material of European and Indian and African blood worked by centuries of hardship. They lived in a land of thorns. Men wore leather to defend themselves from cacti like the *mandacaru,* which flowers only at night and symbolizes resistance to the harshness of the sun. When desperate from thirst and hunger they shared the cactus with their cattle. The yellow, berrylike *murici* they breaded with flour and soaked in *cachaça,* the rind of the *xique-xique* cactus was baked and sweetened with molasses, and the roots of the *umbuzeiro* tree dug up, tapped and drunk. Palm cactus still form the fences to all the little houses in the *sertão,* a hedge in both senses of the word. Richard

Burton, in one of his innumerable footnotes, says the word is simply a contraction of *desertão,* a "large wild."

As Burton drifted down through the northeast he wrote down his impressions of people on the river, and he was saddened to see life becoming "the dullest affair of unvarying shape and changeless colour." Especially in Bahia: "Everything that we see denotes poverty, meanness and neglect." To him, it was a "torpid semi-civilization." Like a good Victorian, Burton prescribed something similar to government's present plan for the valley: more agriculture. The list of crops Burton identified or foresaw thriving with a little investment in water was dizzying: hops and grapes, cereals, all the subtropical fruits and vegetables, coffee, tea, tobacco, vanilla, fine hardwoods, drugs and gums, carnauba wax. He calls it "vast and unexploited wealth."

A traveller in northeast Brazil cannot fail to be conscious of all this history. At the simplest level are the buildings deposited by the Portuguese as they fanned out from Salvador, a miraculous reassembling of local materials — tile, rock, trees from the tropical forest — into the forms of the old country. To enter the chapel in Penedo, at the mouth of the São Francisco River, is to step back into a Europe of the seventeenth century, right down to the reliquaries. History is in the colour of people's skins too: on any street in any town from Salvador to Fortaleza you see features and hues that speak of the crammed and fetid slavers, the *droit de seigneur* of the great landowners whose only qualification was good connections to the Crown, the murders and enforced miscegenation of the *bandeirantes.* I have a Brazilian friend, a biologist, whose grandfather firmly believed in — actually *recalled* — the hunting down of what few Indians were left. Statuary commemorates conquest and violence, and even the shameful accomplishments of the *bandeirantes* are memorialized in monuments to Lampião, an especially successful bandit of the nineteenth century.

But there is also redress. The extraordinary photographs of Sebastião Salgado, most famously his image of miners grappling in the mud for a tiny piece of gold, have ever since been linked with the *Movimento Sem Terra* (MST), the "organization of the landless rural workers" that agitates for land reform in Brazil. Many of Salgado's early photographs, the ones that made him famous, were taken in the northeast and have become emblematic of the struggle to survive: faces with the features of rural Italy or Portugal seen through a genetic kaleidoscope of ten generations of mixing with Indian and African and burned leathery by the sun; caved chests above pants held up with string on a frame that might be forty years old or seventy.

The official path of land reform in Brazil is registration of land that's already occupied, but it's slow and expensive and spectacularly inefficient. A story I heard in Pernambuco is typical: in the early sixties, a rancher gave three thousand hectares of his land to farmers whose families had worked it for generations. They divided it up, but few could afford to register it. Then the entire thing was sold, then sold again, this time to the World Bank, who set about ensuring titles. Only two hundred hectares were actually completed — regimes change with governments — and finally in 1993 government hired more people to go out and measure all the holdings. So far, that's all the farmers have after twenty years: a description of "their" land, but still no title. Even getting to this point has been a constant struggle against attitudes entrenched for centuries. What seemed a simple gesture — "giving" the land to the people who had occupied it for so long — turned out to be impossibly complicated.

At the grassroots level, agrarian reform began in earnest in the early sixties, despite persecution during the dark years of the military dictatorship. First to emerge was the Pastoral Land Commission (CPT) in 1975, and the CPT is still a force. The *Movimento Sem Terra* is one of the more recent of movements and is the most successful, with its own flag and anthem, although with 46 per cent of the land in Brazil still owned by 1 per cent of the population, there's still plenty of room for progress.

The MST was formed in the mid-1980s, with the slogan "Occupation is the only solution." Government reforms struggled to catch up, managing at first to recognize settlements for only ninety families. The election of Collor de Melo as President of the Republic in 1994 resulted in repression and killings, and the MST became more belligerent: "Occupy, Resist, Produce" became the new slogan. The new government of Fernando Henrique Cardoso emphasized export-oriented agriculture, and the MST responded with "Agrarian reform, struggle for all." Two years later came the massacre at Carajás, in which state police in Pará killed twenty-one landless people. Today the slogan is "Toward a Brazil without *Latifundia,*" and the list of official settlements — the next stage after a group of families actually occupies a parcel of land — is slowly growing. But violence continues: in 2004, five landless workers were murdered by fifteen gunmen on the Terra Prometida camp; the owners of the Fazenda de Nova Alegria (New Joy Farm) were suspected.

Control of land meant control of water. In Canada, the very idea of water being under the control of anyone but government (which responds, at least in theory, to the wishes of the people) is still anathema. As with much of the world where water is plentiful, you raise the issue in Canada at your peril. But in the Brazilian northeast, water has always been privatized — not officially of course, where it enjoys the theoretical status of a legislated human right — but in hard, everyday reality. Small dams were built throughout the big *fazendas,* blocking rivers when they filled temporarily in the rainy season. But the resulting reservoirs or *açudes* were for animals first; people had to share, and they needed the permission of the owner. Water became property.

If you survive by subsistence agriculture, you're the first to suffer when water dries up. It's convenient to flip this around and say that "poverty is the result of the lack of water," but there's more to it than that. If people in the dry parts of the São Francisco basin don't own their land, grow just enough food to survive and don't have any way of accumulating wealth, that in itself might be a pretty good reason for being poor. If you rely on rain-fed agriculture, then having formal property rights is obviously a plus. Another way of saying this (and critics of water development in the northeast have been saying it for years) is that water scarcity hits hardest at excluded social groups, transforming a natural phenomenon into a human disaster. If you're the landlord and the pipe is built on your land, you expect to get most of the water.

WHAT HAPPENS TO THE RAIN?

The biggest puzzle is that, really, conditions shouldn't be so bad in the northeast. First of all, there's nothing especially inhospitable about semi-arid regions. They occupy 55 per cent of the world's agricultural lands. Yes, the Brazilian semi-arid region is one of the biggest in the world, the size of France and Germany together, and with eighteen million people it's well populated. But it's also one of the wettest semi-arid regions in the world, with an average of 750 millimetres of rain falling every year, plenty of water for *any* part of the world. The problem, or the first of the problems, is what happens to all this rain.

In the words of Celso Furtado, the great commentator on the northeast, "it falls and it runs." The soil is thin; throughout the vast majority of the northeast, the rock beneath the soil is crystalline and near-impervious, so most of the water just slides away. Given the closeness to the equator and the endless days of sun and wind, it simply burns off (an evaporation to rainfall ratio of three to one, to be exact). Finally and most cruelly, that comforting 750 millimetres of rain is capricious. Some places get more than others; it's literally unpredictable. There *is* a rainy season, four or five months of it, but it might not even happen in some years. Even when it does it can be after a stretch of seven months with almost no rain at all.

If you look at charts of monthly precipitation over a span of a few decades for one of the driest places — Juazeiro, say, in Bahia, one of The Counsellor's many stops — it's a planner's nightmare. The only constant is that it doesn't rain much in May to October, but the rest might as well be random. Some years, rainfall is nearly flat, month to month; some years it spikes in December, some years in March. One year might get as much as 1,000 millimetres; another, only 185. Where a graph of annual rainfall in Berlin looks like a row of perfectly capped teeth, the one for Juazeiro looks like the mouth of a jack-o-lantern. Yet both cities have the same total rainfall. "Irregularity in time and space," the scientific papers call it. It reminds me of Brazilian society which, with some exceptions, tends to be rather fluid about schedules and appointments. Meetings may not start until hours after the scheduled time, but they can just as easily persist through dinner and into the night. It's all taken in stride; maybe a history of being held hostage to an unreliable natural phenomenon has something to do with it.

Nevertheless, life flourishes in the northeast. The plants bears little resemblance to temperate vegetation, and none whatever to the kinds of crops that can only grow with irrigation, but the area is full of lessons in how to survive. In the *caatinga* ecosystem, trees are leafy but not dense, and often spiny. Cacti are common. When the rains finish, the leaves fall to reduce evaporation and the plants turn bleached and ghostly ("white woods"). All have developed elaborate systems for storing water. The *jaú* is a tree that manages to stay green all year by sinking long roots into the soil; the *umba* actually flowers in the dry season and stores water in subterranean reservoirs the size of basketballs. *Vaqueiros* (herdsmen) dug them up for their water. If a tree can do this — survive, even flower and fruit, in the midst of drought — why can't humans?

In fact, they can. There is every evidence that, before colonization, the Brazilian semi-arid was occupied by indigenous people. So the answer is to be found not in the evolutionary superiority of trees over humans, but in the peculiar history of colonization, of what happened after the Indians were pushed aside. History again: the ownership of land and the erasure of indigenous people who had managed somehow to live with the exasperating irregularity of the rains.

Why can't the existing water go farther? There are several reasons. One is simple lack of access (or, put another way, terrible distribution). A good example is the huge Poço da Cruz *açude* in Pernambuco. The reservoir was built in the 1940s and includes a system of irrigation canals — but these have been dry since 1994 due to lack of maintenance. Families who have settled in the area know there's five hundred million cubic metres of water within twenty kilometres, but they can't use it. Such people can't be expected to see the point of spending billions on a water diversion project when all that's needed is to fix the existing delivery system.

Another example: in 1998, government money built the Redenção Canal near the town of Pombal, Paraíba, the place where Antônio Conselheiro began the wanderings that would end with his head going to Rio in a box, for phrenological measurements.

Drought and soccer in Pernambuco, Brazil.

The canal cost 105 million reais, extends thirty-seven kilometres, and is still dry because government decided to redirect water to a different irrigation project. Another 55 million reais was spent in pumps, tubes and auxiliary reservoirs. In the interim, infrastructure along the length of the canal was vandalized and stolen, and had to be rebuilt (another 11 million reais). The proposed diversion project for the São Francisco ignores the lot.

Over the years, the big development banks (World Bank, Inter-American Development Bank, Global Environment Facility) have spent millions on interventions to move water around, but many of these have been left unfinished and some, like irrigation projects that have environmental consequences, may even make things worse. Separate projects are rarely coordinated. This is how bad things have become: the World Bank has made it clear it won't support the São Francisco Diversion project, because it's proceeding in isolation from other projects aimed at redistributing water that's already stored.

Redistribution of water hasn't worked because the conflicts over who gets to use it are so persistent. Some of these conflicts are unsurprising and generic and they're limited to the region surrounding the mainstem river: irrigation, hydroelectric power, flood control, water for industry and human consumption, navigation. Competition between these users happens on most major rivers. Navigation is especially interesting, because it used to be a big thing on the São Francisco. Two sections of the river have historically been navigable: the thirteen hundred kilometres between Pirapora and Juazeiro, and the last two hundred kilometres between Piranhas and the sea. Richard Burton obviously thought a lot about this aspect of the river. He envisaged a railway from Rio de Janeiro that connected with steam vessels on the São Francisco: "Evidently much reform is here wanted, and it will come in the form of a steamer." He was wrong about the railways — in Brazil you rarely see a train, because the road-builders were quicker off the mark and had the better political connections — but part of his vision came to pass. For eighty years the steamer *Benjamin Guimarães* went up and down between Pirapora and Juazeiro. Now, large boat traffic has been squeezed into less and less of the river as it's become chopped up by dams and silted up where flow has dropped too much. The grand old paddle-wheeler is only fired up on special days or for film crews.

Asking God for Help

How did *nordestinos* deal with the iniquities of water shortage, including those imposed by nature and those inflicted by their fellow man? Revolt and reform were important, but there was a third "R": religion. You cannot attempt an explanation of Brazilian water politics through examining cultural history alone. In Brazil, culture and religion are so closely ingrown that to try to consider each separately is like trying to remove the orchid that blooms in the crotch of a tropical tree: the roots of one penetrate too deeply into the heart of the other.

For the owners of big *latifundia,* when cattle suffered in time of drought, it was a tragedy that outweighed any privations of the humans who worked the land. For that level of suffering, it was convenient to cultivate a fatalistic mentality, something that was implanted by the Church and that you can still see every day in the *sertão*. Fatalism made drought into the work of a vengeful God, a sprinkling of holy water where it suited Him, a stern withholding where it didn't. In the popular imagination, drought was explainable this way: it was God's punishment, and the difference between rain and no rain was faith. Even worse, to interfere in this cruel scheme of things was to interfere with divine will. Water could be prayed for, but storing it was a delicate matter. Catch-22.

So access to water historically required loyalty to the landowner, with a healthy dose of prayer. The custom of "kidnapping the saint" — stealing the image from the church in times of drought and not returning it until it rained — became common. (As long as you believed the saint spent every day of captivity working hard to intercede with God, results were guaranteed. Eventually it rained, and the saint, having performed, was returned triumphantly home.) Less likely to work was the custom of lighting candles in dry wells. Another was the practice, not unlike Groundhog Day in North America, in which a block of salt was placed on a table on the nineteenth of March, St. John's Day; if it dissolved, that meant rain would follow. Humidity may have had something to do with that one, but then again, consulting my table of historical monthly rainfall in Juazeiro, I see that, more often than not, March shows a hefty spike. In 1997, Juazeiro got 422 millimetres in March.

All of these customs have in common the idea that humans are helpless to do anything themselves, that the best they can manage is to politely remind God they're dying of thirst. It's an attitude not easily changed. Critics of land dependency in Brazil are unsparing when it comes to the role of the Church; the image of God the punisher, they say, is no different from that of the cruel colonial landlord. Nevertheless, NGOs pushing land and water reform in Brazil have always to respect the centuries-old pull of fatalism, delicately inserting the idea of human responsibility into a new conception of managing a "gift from God." When Barbara Tuchman told the story of Montezuma's fall at the hands of Cortes, she had this to say: "One cannot quarrel with religious beliefs, especially of a strange, remote,

The water truck in Brejo da Madre de Deus, Pernambuco.

half-understood culture. But when the beliefs become a delusion maintained against natural evidence to the point of losing the independence of a people, they may fairly be called folly." She might just as well have been writing of the fatalism of the northeast and its result, the *nordestinos'* acceptance of their dependence.

Stored water in the northeast is still, by and large, available only through the good graces of someone earthly, someone with whom you have no contract, only a historical arrangement of favours for fealty. The hated *carro pipa* or water tanker has become the symbol of that dependency, and you see them in every town in the northeast. When they're not making deliveries or filling up at a controlled reservoir you'll find them parked outside the town hall, because they're controlled politically. *Agua,* it says in big letters on the back, along with the phone number everyone knows.

A vote for the incumbent mayor helps keep you on the list; at election time, the landowner delivers a block of votes, and to go against the flow of votes could easily reduce the flow of water. *Nordestinos* have thus always lived in two parallel universes: the day-to-day religious one, where life goes on with work, dances, festivals, marriages, births and deaths, and the political one of favours and fealty and elections. Both have in common the acceptance of powerlessness; neither contains a shred of the responsibilities and rights of something Brazilians claim to hold dear, and that my policeman friend Arley reminded me of after I was robbed on the beach — citizenship.

THE DROUGHT INDUSTRY

There are many ways to deal with water shortages in a place like the Brazilian northeast; the one you choose to invest in depends on who is going to get the water. Water for poor families in the *sertão*, the backlands, is one thing; water for large-scale agriculture for the export market is another. Desalination is expensive, at least ten times the price of groundwater or river water, and it's not widely used. Large reservoirs have been the preferred solution so far, but they're expensive and they don't serve small communities very well (they're usually government-controlled, which leaves access vulnerable to politics.) Smaller reservoirs need smaller dams and they can be community-built and managed so that water is available to more people over a wider area. But all reservoirs suffer staggering losses from evaporation. Evaporation is so fast in northeast Brazil that a cow dropping dead of thirst doesn't even have time to rot. The parts the *urubu* don't consume stay in place like a faded leather jacket stretched over a row of chairs.

If you want to get water to the poor, especially where they're spread out over the land, storing rain in concrete cisterns or tapping groundwater are the ways to go. Cisterns are cheap, and the existing groundwater is adequate in some areas for local families and small communities; it's already been mapped. Streambed collectors are also simple and cheap. Cisterns and small dams have been used for centuries in Brazil's northeast, since the time of colonization. In India, rainwater collection and storage goes back 5,000 years, using the same simple methods you can still see today on a drive through Paraíba or Ceará: small dams, storage tanks or cisterns, with simple pumps thrown in as the technology became affordable. All of these solutions are decentralized and small-yield, so they make sense for spread-out users, and they don't have many environmental effects.

Large-scale water diversions are just the opposite. The northeast corner of Brazil is the kind of place that the American southwest used to be before people decided to "make the desert bloom." In America, the lack of water in one place was simply seen as a challenge to find it someplace else, and the result was the spectacular folly of the Colorado River diversion. Other counties have done it too: in China, the Yellow River is now a trickle as it approaches the sea; in Egypt it's the Nile; in India the Ganges.

Politicians and engineers love water diversions. First and foremost, they cost lots of money, and there are few administrations without deeply ingrained connections with the companies that build things. Second, they are spectacular evidence of action, another thing governments need for their survival. Best of all, they do the impossible: they bring life-giving water to thirsty places. That's a powerful argument, one that goes straight to the heart. "Vote for me and drink" — in parts of the Brazilian northeast, in the São Francisco basin, it still works that way. It's called the *indústria de seca,* the "drought industry."

The misery of *nordestinos,* trapped and thirsty on land they don't own, has been a staple of political rhetoric since the time of Dom Pedro II (whose solution wasn't much different from today's, namely, bring water from somewhere else). In the 1950s and '60s, agencies created especially for development of the northeast ran into a military coup and ended up promoting industrialization and a government-supported exodus to the "promised land" of the Amazon. Both strategies backfired; both contributed to migrations in the opposite direction, to the *favelas* that grew up around the major coastal cities. Nor do massive irrigation projects help poor people. They simply export water in the form of fruits and vegetables.

And now there is a larger conflict looming, and it goes beyond what a person flying over the river would reasonably think of as its sphere of influence — the farms it could irrigate, the factories it could cool, the dams it could tolerate. Picture that limb again, the valley of the São Francisco, raised and shaken at the Atlantic. Well, the plan now is to stick a needle into the biggest artery and drain some of the blood completely away, to a patient whose disease is so grave, the doctors say, that only a drastic measure like this will save the day. The disease is poverty; the place is the northeast, and the goal is "Water for those who have none." Welcome to the Project for the Integration of the São Francisco. The Diversion.

CHAPTER 13

HIS EXCELLENCY'S THING

Machismo and the engineering of rivers

Who will benefit from the São Francisco diversion? Roadside sign in Paraíba, northeast Brazil.

In the northeast, water is the oldest issue of all. "Water for those who have none" is a good catchphrase for the latest megaproject, but it's a cruel, even insulting simplification of a long historical process. It's almost . . . funny.

Brazilians like their jokes. They like them dark, and they like them dirty. Typically dark is the one about the founder of TAM, the domestic airline. The man died in a plane crash in the 1980s, near the city airport in São Paulo, so people would hum *Tam, Tam ta Tam* to the tune of Chopin's funeral march. The "Tam" sound in Brazilian Portuguese is very nasal and resonant; it really does sound quite dirgelike. Someone told me this joke in the Brasilia airport, as our flight was being called.

Maybe the dirty ones are less cringe-inducing. I heard my first dick joke at a party in a peculiar company town called Furnas. Furnas is also the name of the company, one of several big hydroelectric outfits operating in Minas Gerais. This one was on the Grande River (the same river where I first saw those trapped *curimatá*, at Volta Grande dam) and I don't remember much about it beyond a big hatchery and a lot of fish ponds dug into the earth for producing fingerlings to stock the reservoir — that Japanese model again. In the evening, the sweet smell of eucalyptus floated through the town. The party filled a cafeteria building attached to a string of dorms the company maintained for researchers and university students. There was a swimming pool outside, and I walked out there several times in the course of the evening to stand beside the dark water, listening to music and laughter and looking up at blackness beyond the arc lights. Around the corner an outdoor barbecue was set up next to plastic tubs of bleeding meat and somebody's car with the doors flung open, a stereo thumping into the night. Trees sizzled with cicadas. It was one of those nights where, as Burton said of a gathering in the colonial mining town of Diamantina not far away, "supper never seemed to end."

Inside, Hugo Godinho had just finished telling the story of how he lost his wedding ring early in his career as a veterinarian, palpating a cow's ovaries. "In those days we didn't use gloves," he laughed. "It was awfully slippery in there." I was still thinking about that ring swimming around in the cow's uterus when the dick joke came along. It was one of a string of jokes being told by two people at my table: a small Japanese-Brazilian woman and a lanky biologist named Weber who was clearly in the wrong field. Weber, I wanted to tell him, forget your plans for becoming a biological consultant and get into stand-up comedy. Weber had all the tools. He waggled his eyebrows; he did a great double take, jerking back in his chair and bugging his eyes; he dragged out the pause before the knockout punch. As Weber and the Japanese woman went back and forth, the jokes escalated in complexity and filthiness. He had already warmed up with a few "Portuguese jokes," the Brazilian equivalent of the "Polish joke" in the U.S. or the Canadian "Newfie joke."

First Portuguese joke: Two Portuguese guys are on an escalator. It stops. One says to the other, "Hey, it's got steps, we can sit down!"

Another Portuguese joke: During the First World War, Germans eliminate the Portuguese in the trenches by calling out "Joaquim!" and then "Manoel!" One by one the Portuguese stand up and are picked off. Finally there's only one left, the smart one. A voice comes out from the Portuguese trench: "Fritz!" All the Germans stand up and shout, "No Fritzes here!" The last Portuguese yells back, "Ah, but if there *were* . . ."

The dick joke, when it came, must have been Weber's specialty and to be honest I've forgotten the punchline, but that doesn't matter.

"So the guy stood up," Weber said, his eyes like saucers, "and *whump.*" Weber leaped to his feet, swept a plate and some empty

Antarcticas aside, and clutched his groin. Then he raised his hand to shoulder height, swung it and slammed the imaginary penis onto the table. It was a very big penis. In this version of the joke there were actually several more penises because, like so many Brazilian jokes, this one had the kind of racial overtones you can only get away with in a society that is genetically jumbled. The key thing was that *whump,* the dick slammed on the table; every time Weber did it the Brazilians whooped harder. Whatever the gesture meant, it was nothing new to them. I didn't understand the joke at all, but I knew a dick when I saw one. Weber and his dicks were pretty funny.

Years later, a friend told me something even better than Weber's joke: that the dick-whumping thing actually happened in real life. Erika is tough and outspoken and has a sudden, melting smile so big the crinkles at the corners of her eyes shoot right back under her hairline. Erika originally trained as an architect. Now she manages community development projects, the last of which saw her Brazilian counterpart shot to death by political rivals. We were sitting in her office at the University of British Columbia. Trees moved outside the window and Schubert murmured from her computer.

"This really happened, when I was a student. The dick thing. We had one professor, the guy was brilliant but he was a mess. Always drunk, or taking dope. After the thing happened, the university had to get rid of him."

"The thing with the dick?" I asked.

"Really. Totally a Brazilian thing. He was at a faculty meeting, he was stoned, he got into a big argument and then suddenly, *whump.*" Erika grinned and did a good imitation of Weber, swinging the imaginary organ like a lump of bread dough. "That was it for him."

"So, the plan to divert the São Francisco," I said, "that's someone's dick, right?"

"*Oh* yes," said Erika.

Putting in a big tap

Water diversions can be as small and straightforward as extracting water from a river to pour directly onto neighbouring fields — the kind of thing seen from your car as you drive through a river valley or, even better, from the air, the river snaking through an arid valley clothed in a coat of startling green. These kinds of diversion are cumulative, a bit here, a bit there, and can add up over the years to an extraction of so much water the river starts to falter. Flow rates slip, temperatures creep up, sediments get dropped earlier, and what was once a vigorous corridor for aquatic life and the terrestrial animals that fed on it just isn't livable anymore.

But what happens if you want to bring water, not a few kilometres from a river that's already there, but *hundreds* of kilometres, to a place that has no water at all? Compared to this, direct extraction of water for nearby agriculture is nothing. Now you're talking canals and aqueducts that run off over the horizon, massive pumps to lift the water over mountains, new dams to store it behind. Concrete! Trucks and excavators! Armies of workers! Moving water from one geographic basin to another, now *that's* a challenge!

River diversions are expensive. They need a lot of maintenance. Above all, they embody somebody's idea about where water should go. So while diversions can theoretically provide "water to all," in practice they tend to benefit large agricultural producers the most. Big diversions are, after all, megaprojects, managed by governments or corporations, and the natural beneficiaries are those able to lobby for the water. And one more thing: diversions can actually waste water. The ones that involve vast systems of open canals still send water flying off into the atmosphere at record rates. Large-scale water diversions are not the only examples of megaprojects with folly written all over them, but they're pretty good ones.

If you look at the big global water diversion projects, most of them refer back to the Colorado. Proponents of water diversions use the Colorado to show what miracles can be accomplished with dynamite and concrete; opponents point trembling fingers and say, "Look what happened on the Colorado!" The Colorado River is about seven hundred kilometres shorter than the São Francisco, but there are enough similarities to make some useful comparisons. The longstanding ties between the U.S. Army Corps of Engineers and the politicians and engineers behind the planned São Francisco diversion make comparisons imperative.

When the Spanish first arrived where the Colorado empties into the Gulf of Mexico, the delta received ten times the water it does today. Most of it now goes to farms and cities in the United States. The Hoover Dam, for example, is one of three major dams that make it possible for Las Vegas to have green lawns and choreographed fountains, and a system of canals redirects water across the Sonoran desert and through two mountain ranges, as far west as San Diego. The Imperial Valley in the Colorado Desert, together with Mexican farms on the other side of the border, sucks up most of the river, creating an improbable two hundred thousand hectares in produce and feedlots. The Colorado diversion even created a foul and salty lake, the Salton Sea, when a canal blew out in 1905 and it took two years to plug the hole.

The environmental cost of all those apricots and oranges is steep, and the politics of "sharing" the water between seven states and two countries are Byzantine. Predictably, states in the upper basin (the "producer" states of Colorado, Utah, Wyoming and New Mexico) resent their water going to the "consumers" (California, Arizona and Nevada), who further bicker and deal-make amongst themselves, with California historically taking over chunks of its neighbours' allotments. Arguments rage, and the language is in cubic metres per second as the river and its feeder tributaries are measured out to within a drop of their lives.

Thirsty regions, if they have money, make deals for more water, like the recent one that has the Imperial Valley making a handsome profit by selling surplus water to San Diego, thereby further draining the already putrid Salton Sea, concentrating its agricultural pollutants and beaching truckloads of tilapia to rot in the sun. Moviegoers old enough to remember *Chinatown* know something of the chicanery of Los Angeles water politics during the 1930s, although they probably don't realize that the attendant draining of the Owens Valley led to Owens Lake's drying out and becoming the biggest single source of air particle pollution in the country because clouds of toxic agricultural dust blew off its caked surface. Technical "solutions" to problems like these tend to be ridiculously expensive, like the $1 billion desalinization plant built by the United States to remove all the salts the Colorado collects from agricultural runoff before Mexico gets its allotted 10 per cent of the river. But what's $1 billion when you've already spent four times that on just part of the diversion system, the Central Arizona Project, the largest and most expensive water transfer ever built?

These are head-spinning numbers. Once a country embarks on a project like the Colorado diversion, there's no turning back, and there has to be a pretty big payoff beyond the boom in construction and votes for the people promoting it. The answer, of course, is as old as the first aqueduct that drew water out of the Ganges five thousand years ago: food. Not farming on a small scale, the way it used to be, but on the scale of the Imperial Valley in California or the fields of engineered soybeans you always seem to find yourself driving through in Brazil. Bigger is better. And Big Water = Big Agriculture is the equation that matters.

AN OLD MODEL WITH A NEW PAINT JOB

The idea of diverting water from the São Francisco River to drier, more northerly river basins is an old one. People have been exhuming it for centuries. It has enormous political appeal because the logic (at least at the level of political rhetoric, which is a pretty low level) is unassailable: there's drought, people are dirt poor and suffering, bring water into their area and you help them. "Water for people" is a powerful mantra, especially when combined with images of children balancing pails on their heads, donkeys pulling ramshackle carts with rusty tanks, emaciated zebu cattle standing in riverbeds crazed by drought.

Around 2004, as the diversion project actually appeared to be staggering past the starting line, the problem for its proponents was the age of the idea. Twenty years ago it might have sneaked by. But in the decades since it was first put forward, and especially in the previous ten years or so, the arguments of those opposing the idea had become more sophisticated. People had done their homework, and the results of other large-scale water transfers had been studied hard. Much of this work was done by NGOs.

I don't know whether it's a delayed response to the flowering of democratic government in Brazil or just natural argumentativeness, but the non-governmental organization in Brazil — the NGO — is alive, well and growing. The gathering of nations at the Earth Summit in Rio in 1992, the first such global meeting where NGOs were a force to be reckoned with, must have sent a powerful signal to Brazilians, because more than half of the existing NGOs in Brazil sprang up since then. In all, there are more than 275,000 Brazilian non-profits working strenuously for one cause or another, with rural producers and rights organizations concentrated in the northeast. Most of the NGOs are small (in Brazil they're called "ongies," ONGs) but some have the wherewithal to employ researchers and experts and even, in a weird Brazilian way, to get most of their funding from the government they attack.

So the present government's project for diverting the São Francisco has been gone over like the used car it is. Academics, NGOs and the media have chipped at the shiny new paint to reveal layers of primer and body filler, inspected the oil and found telltale streaks of water, tested the failing compression and kicked the cheap recapped tires hard enough to delaminate them. These are no ordinary car buyers, no young couple wandering onto the car lot on a fine spring day when the plastic pennants snap in the warm breeze and the salesman bounding out of his shack has an extra spring in his step. These buyers have done their homework in darkened rooms with coffee-ringed reports and testimonials and analyses piled three deep. They walk onto the car lot not with a stroller and gurgling firstborn but with an armload of checklists and guidebooks. I imagine the salesman looking up from his coffee.

"Um, Dave?"

His colleague lowers the racing form. "Bob. Old buddy." Dave stretches and begins to rise, but Bob jumps to his feet in front of him, blocking the view of the cars outside. Behind him a dour couple is approaching the Diversion Roadster Mark IX, dressed in brown, wearing steel-toed boots and determined expressions. Bob gulps.

"Dave, heh-heh, just remembered. Stupid of me. The old lady, promised I'd pick her up right about" — he consults his imitation Rolex — "Lord, right about now." He ducks suddenly out the back door, calling "I owe you one, pal," and hurries away past a gleaming

Toyota Corolla with three hundred thousand kilometres on it.

The used car that is the São Francisco Diversion Project has had a few owners. There were two other "modern" versions: the first, in 1985, proposed withdrawing an impossible three hundred cubic metres per second to the Castanhão reservoir; the second, in 2000, proposed withdrawing forty-eight cubic metres per second in two canals. The shiny 2006 model aims at twenty-six cubic metres per second, a number freighted with distrust and multiple interpretations, as we will see. Water will be taken out of the river at two points, both of them in the state of Pernambuco. Water withdrawn near the city of Cabrobó will be diverted north, through four hundred kilometres of new canals as well as by way of existing rivers, ending up in thirteen large reservoirs. The water taken from behind the Itaparica dam (the other out-take point) will go roughly east through its own new network of two hundred kilometres of canals, and it too will eventually merge with several independent river systems.

Getting all that canalized water to the distant reservoirs means building pumping stations, aqueducts and tunnels, because this may be dry country but it's not flat. A cross-section of the two wings of the project looks like an escalator that can't make up its mind: up it goes, then down, then up again. There are hundreds of highways to cross or dive beneath, and at the end of the escalator wait rivers — the Jaguaribe, the Apodi, the Paraíba, the Piranhas-Açu — that are total strangers to the São Francisco. It's an arranged marriage of watersheds, loveless and ill-advised; the proud parents are engineers and politicians, not biologists.

What would induce people to buy such a project? What's the track record of big diversions like this? There have been lots of other large-scale water diversions elsewhere in the world, and critics of the São Francisco project love to point out what's gone wrong. The Chavimochic Project in Peru pulled water out of the Santos River and resulted in salty soils and difficulty covering operating costs through water payments. In Spain, the Tajo-Segura aqueduct also resulted in soil salinization. But the astonishing constructions of the thirties and forties in the Colorado Basin (the very projects that fired Brazilian determination to create its own diversion project) are always at the top of the critics' list. Besides the obvious and indisputable effects on river flow and biodiversity and the intrusion of saline water into water tables, there are fascinating social parallels. Bringing water to unused arid lands in the Imperial Valley in California only helped solidify the hold of a few rich families, and land values skyrocketed. When water was taken from the Owens Valley to feed the city of Los Angeles, water politics resulted in an ecological disaster for the donor area. And the interminable wrangles over the Colorado Compact, which divided up the waters of the Colorado between donor and receiver states, should give any Brazilian governor pause.

Government promotion of this gargantua is skilled and extensive. Much is made of the historic allure of Velho Chico. I saw a stunning exhibit of photographs of the valley in an upscale shopping mall atrium in Recife, expensively blown-up images of *nordestinos,* Indians, fields of grapes and soybeans, the tiny barnacle of the dam at Delmiro Gouveia near Paulo Afonso. The exhibit was paid for by CHESF, the Electric Company of the São Francisco.

The Brazilian government also produces various information packages on its diversion project, as governments must. One of them features a fine reproduction of the original 1852 engineering survey of the river, which lends historical gravitas even if the forest depicted by the long-dead engraver is long gone too. A glossy thirty-page booklet on the transposition has a weird collage of images on the front cover: a farmer, some oranges, a young girl drawing water from a tank, a smocked figure that could be a child or a wizened old lady pouring water into a shiny stainless steel tub. Inside, engineering diagrams show a slice of the watershed in elevation: the river leaps over mountains, crosses highways, flows downhill in canals and aqueducts, rests in reservoirs new and old. Laid out like this the project is truly staggering and the centrepiece, a three-dimensional view from the Sobradinho reservoir looking toward the Atlantic and showing the canals and pipes radiating out across the northeast like the veins of a leaf, makes you realize why critics call it "pharaonic."

My favourite, though, is the booklet for kids — at least I hope it's for kids. This one features a cartoon family and all the questions are asked by the little boy and his toothless grandfather, who spends most of the book standing off to the side, scratching his head and sending up clouds of question marks. Mom tends the goats and the washing; Dad, who seems to represent reason and the Brazilian federal government, grins incessantly and dispenses answers. In the end, he sets off across the arid landscape toward a construction crew on the horizon, clutching a lunch pail. The final page shows a map of the northeast with a big tap stuck in the middle and the family waving happily.

ROGUE'S GALLERY

Proponents also like to cite examples of other large-scale water diversions, usually starting with subterranean canals (*qanats*) in Mesopotamia and the aqueducts of Rome. Neither is technically a diversion into another river basin, but the comparison with ancient civilizations lends an imperial touch. In Brazil, the two major existing diversions are relatively small ones that provide municipal water to Rio de Janeiro (from the Paraíba do Sul by way of the Guandu River) and São Paulo. There's also the unofficial one that's served Belo Horizonte for thirty years, in which water from the Paraopeba River goes into the city and emerges, full of sewage and contaminants, into the Das Velhas River, where Richard Burton began his journey. It's something of a dirty secret because the Das Velhas has almost been killed by that one, and it just happens to flow north to connect with the São Francisco.

In present-day Spain, the same issues that plague the São Francisco torment the Ebro, where planned diversions will flow through over eight hundred kilometres of new canals to reduce water shortages in the dry south of the country. In Nepal, the Melamchi Project will divert water to residents of the Kathmandu Valley by way of a tunnel through the mountains. The grandest scheme comes, as usual, from China, where rivers seem fated to be dammed, drawn down and diverted on a scale that makes the Colorado or the São Francisco seem like minor plumbing jobs. As I write this, the Three Gorges dam inches steadily higher, but there is already a plan to link the new reservoir (biggest in the universe, etc., etc.) with a system of three north–south canals that will ulti-mately create a single network between China's four major rivers: Yangtze, Yellow, Huaihe and Haihe.

You really have to stop and take a big breath to get your mind around this one. The four rivers in question have the annoying char-acteristic of flowing from west to east, separated by mountain ranges. But mountains can be moved. The aim of this spectacular engineering project, which will, among other things, cut through the Qinghai–Tibet Plateau at elevations up to five thousand metres, is to provide more water for the industrialized north (in this case, industry provides a much better return per litre than does agricul-ture). And just in time, too, because so much water is being pumped out of the ground that the water table there is dropping three metres every year. You can almost hear the sucking sound. It will cost over $US 50 billion, ten times the cost of the São Francisco diversion.

There are even water diversions that provide comic relief. The North American Water and Power Alliance (NAWAPA) and the Great Replenishment and Northern Development (GRAND) Canal have both been seriously proposed to transfer "surplus" Canadian water to the United States. Plans for NAWAPA first emerged in 1964; the idea was to divert Canadian and Alaskan waters to the U.S. and the northern states of Mexico. It involved damming the Yukon, Skeena, Fraser, Peace and Columbia rivers and actually flooding the Rocky Mountain Trench to create an enormous reservoir. Eleven major BC rivers would be reversed into the Rocky Mountain Trench. The out-flow would end up in Flathead Lake, Montana Lake and Lake Mead, and the new Great Canadian Lakes Waterway would stabilize Great Lake fluctuations and open up shipping routes from west to east.

This insane idea slept until the late 1980s, when it was reintro-duced in the United States. BC's Premier at the time, Bill Bennett, noted that fresh water going into the ocean was "a waste" — an argument politicians still use in Brazil. (People trying to make a profit out of water generally prefer to ignore the water cycle: Snowcap Waters, a bottling company in Union Bay, British Columbia, argued in 1992 that allowing 294 billion gallons of fresh water to escape into the Pacific from British Columbia every day was "sheer waste.")

Then there's GRAND. The GRAND Canal has been called the "eastern

NAWAPA" and should forever lay to rest the theory that engineers and politicians are not, at heart, creative people. This one reads like the stuff of late nights with mind-altering drugs. Dating from the sixties by way of several Canadian politicians including the dam-building former Quebec Premier Robert Bourassa, the project was to dam the mouth of James Bay, somehow edit out the salt water with "sluice gates" and turn it into an enormous lake. Water from the new lake would then be pumped to the Great Lakes. According to Tom Kierans, the first promoter (of course, an engineer) the fundamental problem of Hudson Bay is that there is "too much fresh water." (That waste thing again.) Low salinity levels meant low productivity, thus no commercial fishery, so what was the point of all that water? Apart from damming James Bay, the plan would require nine inter-basin transfers. Pumps required to regulate the reservoirs were already conveniently available: all you had to do was reverse the turbines on existing hydro dams. In 1994 the capital costs for this scheme were estimated at $100 billion. I think we can safely use the term "pharaonic" for that one.

Neither NAWAPA or GRAND has gotten off the drawing board — yet — but the ideas keep returning like an LSD flashback. And at least the planned alterations to the landscape, however ludicrous, are easily visualized by the average person: canals, reservoirs, pumps, rock blown into the sky and waters boiling into places they weren't meant to be. For people horrified by such images, a major water project is at least a tangible target. But some water projects are completely hidden. The best place to see a really big one may be in Libya, where $40 billion worth of tunnels radiate from aquifers beneath the Sahara Desert to farms and cities in the north. Except that you can't see it at all, because it's happening underground. Out of sight, out of mind.

REVITALIZATION: THE MIRAGE

There's another way the proponents of the São Francisco diversion promote their project — by dangling a carrot. On one fisheries project in Brazil we used to chase a mythical pot of money, spoken for but oddly unspent. It became a kind of game, something to do on a cold Canadian November afternoon when the view outside was of grey clouds and bare trees, when thoughts turned to steaming rain and migratory fish a continent away. The pot had a minder, a senior bureaucrat I'll call Sergio. He was an affable guy, earnest and convincing about releasing the money: the cheque was in the mail. Time and again, though, the final step just didn't get taken: a key functionary was transferred, there was an election, Sergio forgot how to dial the phone.

I eventually met the man, in his office in Brasilia. He embraced me like a brother, offered me *cafezinhos* from the usual silver tray, loaded my bag with glossy Ministry publications I left for the bemused maids in the Hotel Alvorada. Sergio had decent English, as most of the higher bureaucrats do, and his current favourite word was "fantastic." I was fantastic. The project was fantastic. What we — Sergio and I — were contemplating for the river was (and he really said this, coming close and seizing my shoulders) "super-fantastic." But the cheque stayed in the mail.

That money was for "revitalization" of the São Francisco; it's probably still in Sergio's drawer, in a thick wad of virtual reais earmarked for hundreds of specific projects that have been approved, but not released. Brazilians have an expression for projects like these: they've "never left the paper." Revitalization of the São Francisco is usually costed at about a billion reais versus six to twenty billion for the diversion, and it's the sum of all those blindingly obvious, meat and potatoes actions that absolutely everyone, including governments, industry, NGOs, agrees have to happen on the river: reducing pollution and siltation, re-planting and protecting the marginal forests, bringing back the fish and their habitat.

The list of side effects of human activity that revitalization is meant to minimize is a long one. Ranching, for example, erodes soil; deforestation and substitution with irrigated crops or the eucalyptus Barbara Johnsen showed me remove marginal lagoons and simplify the river's course so that it no longer performs its normal ecological functions. Urbanization and irrigation have turned the river and its tributaries into dumping grounds for raw urban sewage, heavy metals, pesticides and fertilizers. Of the 504 municipalities in the basin, only 26 per cent have water storage facilities; only 15 per cent have basic sewage treatment. Critics say, what is the point of with-drawing polluted water and sending it to thirsty people?

And finally, revitalization is needed to counter the effects of all those dams: lack of floods, barriers to fish, siltation. The dam com-panies just can't bear to release water simply to create a little flooded area for fish; they'd sooner cling to the belief that, somehow, migra-tory species can adapt to spawning in reservoirs, as though millions of years of evolution never happened. But flood control is a double-edged sword: flow regulation has resulted in a river that is losing soil in some areas and collecting it (silting up) in others. Islands of sand appear in strange places; sediment eroded from denuded banks set-tles in reservoirs, backing up against the dams and imposing an inevitable limit on their ability to regulate floods and generate elec-tricity. The dams become huge settling ponds and are even frequently hailed as "cleaning the water," a statement that's per-versely true. If you happen to cross the bridge at Três Marias during the rainy season and are lucky enough to witness an opening of the spillway to release some of the buildup, the surface water that thun-ders out is clear and an unnatural blue. When it lands on the muddy river below it appears, for a kilometre or so, as though there are two rivers running side by side, one clean and the other dirty.

Some people even argue that deforestation and dams cancel each other out, at least as far as the effects on sediment go. The reasoning goes like this: deforestation causes riverbanks to erode, yes, but the stuff all settles to the bottom once it hits the next big reservoir. So reservoirs are good. Of course it's not true. The riverbanks stay eroded; upstream of the dams, sediments can bury small towns; reservoirs only trap sand and silt, while the third component of riverbanks, clay, continues downstream to muddy the waters.

Revitalization, meant to attack these ills and more, is the carrot on the end of the diversion stick, funded but forever dangling just out of reach. You have to multiply my own experience with Sergio by thousands, hundreds of thousands, to get an idea of the amount of effort and time spent on meetings and proposal-writing in a country as heavily bureaucratized as Brazil, negotiating compro-mises with traditional foes, all to have projects "approved" and the money never released. On the level of my experience with Sergio it's easily laughed off; the guy's a jerk, what else is new? But in aggregate, when it's spread up and down three thousand kilometres of river, it's political meanness of a very high order.

When Barbara Tuchman defines folly she is, as usual, succinct: "We all know, from unending repetitions of Lord Acton's dictum, that power corrupts. We are less aware that it breeds folly; that the power frequently causes failure to think." She might also have added that majestic construction projects are a favourite way of expressing that folly; the new basilica of St. Peter's in Rome, creation of Pope Julius, has certainly outlived his memory. But Julius paid for it in a way that guaranteed the downfall of the Renaissance papacy: he sold indulgences. The plan to divert the São Francisco is so old and has so many critics it's perfect as a case study for folly. Its long history makes it seem like one of those nightmares where you can see something terrible happening — a car crash, a plane going going down — but when you open your mouth to scream, nothing comes out. Relentless scrutiny has made the central issues pretty clear — a prerequisite, Tuchman would add, for its classification as folly. She would probably also note how the project has been embraced by President Lula, especially after his previously opposing it; such folly she calls "the child of power."

For President Lula, having the São Francisco Diversion as part of his platform didn't prevent him from being re-elected in 2006; it may even have helped. But it didn't stop the criticism, not only because the critics still hate the project, but also because water is vaulting to the top of global conservation issues. Different models for using water — market models, like the one used in the Colorado basin, versus collective negotiation — are hotly debated. In Brazil, water management is a work in progress, a canvas fought over by a small army of contentious artists, the colours riotous one day, erased the next.

In 1997, Brazilians finally got a new Water Law, based on principles like finiteness, economic value, multiple uses, decentralization and people's basic needs. Frail or not, these principles provided a canvas to paint on where before there had just been empty space. The idea behind the new law was to focus on using existing supplies better and more democratically, with less reliance on large-scale water works. Management was suddenly, to use the holiest word in the late twentieth-century bureaucrat's vocabulary, "participatory." River Basin Committees were part of the deal, and of course the São Francisco got its own.

But the shiny new Water Law came too late for the São Francisco, because the river had already been divided up. Every drop was now jealously guarded by what have traditionally been its main clients, the hydroelectric companies. I got my hands on an internal report on water use prepared in 2002 by the National Water Agency, and it captured the official priorities: power generation, flood control, navigation and irrigation. Domestic water supply and the needs of fish were "unable to be considered . . . because of lack of information." So was "dilution of pollution." As far as the report was concerned, there were really only two conflicting uses of water in the São Francisco: electrical power and irrigation. It was clear, right down to the calculated tradeoffs. Each cubic metre allocated to irrigation meant a loss of 2.5 megawatts. The only other caution was to watch out for evaporation loss. Other uses "could be" included in water use models, but they weren't. The report concluded with a recommendation to focus on power generation. If you want some extra water released, you'd better have influence with the power companies. This is the playing field on which the critics of the São Francisco Diversion do battle with its proponents.

WHAT'S WRONG WITH DIVERTING A RIVER?

For river diversions to work, three conditions have to be met. First, the river basin donating the water needs to be an area of low water demand, with little opportunity for further irrigation. This doesn't hold for the São Francisco, where irrigation demand is still growing. Second, the receiving areas need to have plenty of irrigable land; in the case of the São Francisco, the new network of canals will only cover 10 per cent of this land. Finally, the project has to pay for itself. The São Francisco fails here too: apart from the cost of building two thousand kilometres of canals, enormous energy needs to be spent to pump the water 165 metres up. The solution is inevitable: users will have to pay. As one writer put it, "beware of Greeks and their gifts."

By the beginning of 2004, criticism of the diversion was loud and voluminous; as this book was written, in 2008, the volume has only increased. Web sites appeared; hydrologists, sociologists, biologists, activists and clerics attacked the project. Many focused on the stubbornness of government in sticking to an idea that will only perpetuate the unequal distribution of an already adequate supply of water. Again and again, this is the argument that stands out: over the past hundred years, Brazil has already invested in one of the biggest reservoir-building efforts in the world, with more than three hundred of these man-made lakes pock-marking the northeast alone. Why build a completely new system when all you have to do is redistribute that water?

If you look in detail at criticisms of the transposition, most of them are being made by people for whom transfers of water between basins isn't in itself repugnant; what bothers them is that it's simply the wrong thing to do for the northeast, that what really needs to be fixed is how that water is managed. It's not a case of environmentalists clashing with developers; actually, few environmental alarms have been raised. That's not to say such risks don't exist, though. Around two hundred species of fish have been described in the São Francisco, and there are probably another hundred yet to be described. In the receiving river basins, though, there are far fewer species (fifty-three described so far), and almost half of them are endemic (unique to that area). The diversion will inject dozens of new species, mostly as larvae and juveniles, into the new basins, a contravention of Brazilian legislation that forbids introduction of exotic species. It's ironic: the same government that established a National Task Force on the invasive golden mussel is promoting a massive illegal transfer of exotic species within its own borders. If environmentalists wanted a tasty issue to chew on this would be it. But they're not biting. Carlos Bernardo, the biologist from Belo Horizonte who accompanied his compatriots Norberto, Miguel and Roberto on that fact-finding tour to Vancouver, is one of the few public voices on this issue in Brazil. But the fight against the São Francisco diversion isn't really about the environment. Social justice trumps worries about fish.

What *do* the critics say? It's a big project, so a good place to start is politics and money. The project makes some political sense; the appeal of fifty million northeast voters is irresistible. Naturally it depends on just where in the northeast those voters live, and how likely they are to vote: socially, the project is attacked for bypassing the poorest people, providing most of the water for irrigation projects and to coastal cities like Fortaleza where the standard of living is already higher than in most of the São Francisco basin itself. Many argue that each acre irrigated by the project in the northeast simply takes two or three out of the running in the São Francisco basin (an argument that has the tang of Colorado acrimony about it), so politics at the state level are involved as well. Nevertheless, widespread criticism of the project was not enough to keep President Lula from returning to office in 2006, so "water for those who have none" must resonate somewhere in Brazil.

Money is another story. Critics assume the direct financial beneficiaries will include contractors and owners of land next to reservoirs and canals. The federal government has a bad record for completing projects in the region: 180 million hectares of irrigation projects still await funds, and the São Francisco diversion will drain exorbitant amounts of public money. The National Water Agency (ANA) has granted a twenty-year period for the Ministry of National Integration to build the project at a planned cost ranging from $US4.5 billion to $US20 billion. Annual operating costs, to be divided between the receiving states (Ceará, Rio Grande do Norte, Paraíba and Pernambuco) are estimated at $80-100 million. Who will pay for all this?

Not the World Bank: after analyzing the project, the Bank urged investment in local water supply projects and revitalization of the river instead. The Bank is "out," which means that Brazil will do as China did with the Three Gorges Dam: go it alone. Hence the recurrent criticism that bringing water to the northeast at such a cost will end up essentially privatizing it — this time, officially. The cost of water in the northeast will multiply by at least five times, and its actual distribution will be regulated not by the federal government but by a collection of state-owned distribution agencies, each with competing priorities. Because the São Francisco provides 95 per cent of the northeast's electricity, any losses will have to be made up by constructing thermoelectric plants.

Managing all that diverted water will be political hell, just as with the Colorado. Because the historical approach to water scarcity in Brazil has been to build large reservoirs and canals on private land — the infamous *indústria da seca* or "drought industry" — decentralization efforts like the River Basin Committees face an uphill battle against those who have always held the power, and the number of groups squabbling to create a better way is huge: public officials keen on reform, local resource users, NGOs, politicians and an army of technical advisors. Of the twenty-six cubic metres of water that will be taken out of the São Francisco, only five are going to the states of Pernambuco and Rio Grande do Norte, and the rest to the monster Castanhão Reservoir in Ceará. As one critic put it, "whose vegetable garden do you think the rain's going to fall in?"

There will be massive social effects. IBAMA — the Brazilian Institute for Environment and Natural Resources that may be thought of as the regulatory arm of the Ministry of Environment — has not considered the situation of the 34 indigenous communities and 153 *quilombos* (settlements of the descendents of escaped slaves) that will be affected. But these people will be heard from, especially if they are dislocated. The two thousand kilometres of canals carrying water from the São Francisco are designed to distribute water for ten kilometres on either side, creating a watered area that covers only 10 per cent of the semi-arid lands. The geologist Edezio Teixeira asks, "What happens when you run a corridor of gushing water through a semi-arid area? A water stampede happens. And then the people will stay." This migration will be massive, like all the previous social migrations in Brazil's history, and it won't be prepared for. One critic called it "favelization" of the canals. But semi-arid regions are inhospitable. Their carrying capacity — the amount of life that can be sustained — is low. Bringing prosperity to a place like the arid northeast will just overload the ecosystem, and the first thing that will be in short supply will be . . . water.

Seventy per cent of the water to be withdrawn has already been guaranteed to new irrigation projects, so the charge that the diversion is all about "big agriculture" and the imperatives of the global market is persuasive. The "miracle" of the Imperial Valley in California seems to have hypnotized people. You have only to drive through the vast fields of soybeans that cover so much of São Paulo and Paraná, and to know that Brazil is the world's number one producer, to realize just how seductive the idea of "more agriculture" is. The world market for grapes is waiting, so why not make the northeast into wine country? But people living in the São Francisco basin itself argue that there are thirty million hectares waiting to be irrigated in their own backyard, and that so far only 1 per cent of that has been covered. Why give the water away to do the same thing somewhere else?

Then there's the design. Designs have to start somewhere, usually with a good knowledge of the scope of the problem. But with this one there's a big hole in the planning: how do you match water use with availability if you don't even know the real numbers? It's like a budget: you can't plan future spending if you don't know how much you've already spent, and where it all went. Water availability is measured through river flows and groundwater levels, but actual use is much harder to gauge, especially when so much of it is illegal. How do you measure the amount removed for agriculture when it's not charged for? In the northeast, livestock outnumber people in some areas — how much are the animals drinking?

Industry and domestic water use information may be collected but the data aren't centralized, and domestic use is further complicated by being split into actual water supply systems in cities and a large network of wells and small reservoirs in the countryside. In Ceará, for example, only half the population is connected to a water supply, mostly in the coastal city of Fortaleza. And the environment itself is — or should be — a big water user. How much water should you leave in the river so that fish can continue to migrate, so that silt doesn't collect far upstream, so that the mouth doesn't become a pitiful trickle like the Colorado?

The engineering is another big question mark. For a region where evaporation is the real killer, building a vast network of open canals doesn't make sense to many critics. The amount of water presently lost from the two biggest reservoirs on the São Francisco (Sobradinho and Itaparica) is equivalent to one entire Colorado River every year. That's three litres out of every four stored, lost back to the atmosphere, the water cycle making a mockery out of man's attempts to get around it. The seven hundred kilometres of canals themselves will be twenty-five metres wide, five metres deep, and wide open; as a way of getting water efficiently into the atmosphere it can't be beat. And the plan to develop small-scale agriculture along a 2.5-kilometre swath the length of the canals goes against what's known about the soil in the area: it's thin, on top of crystalline rock, useless for irrigation. Building seven hundred kilometres of canals on this kind of surface will be slow, no more than a hundred metres a day, a total of seventeen years of blasting and concrete.

Then there's the numbers game. For those with a mathematical bent, the real bearpit is river flow: how much water is there? How much can you take out? The São Francisco has been officially reborn as a bank account, and managing it is all about doling out cubic metres per second. Everyone has their own numbers, and when you are dealing with water flowing down a river, even a regulated one like the São Francisco, there are so many to choose from (peak, low, median) that you can make almost any argument you want.

The average flow at the mouth normally hovers around 1,850 cubic metres per second and is set by IBAMA at a lower limit of 1,300, to make sure there's enough water for aquatic ecosystems and fisheries. The existing holders of this enormous bank account (the thirteen million people living in the basin) have been allocated 360 cubic metres per second, and they've presently got claim to 335, leaving 25 still available. Of course, the 360 itself is based on assumptions about ecological requirements, which may be way off. But the diversion project contemplates taking out an average of 63.5, rising occasionally to 127. With only 25 available, taking out 63.5 means cutting back on basin use of water.

There's a big difference of opinion here, on whether the amount withdrawn should be based on flow at the mouth or the amount not already spoken for. Proponents who insist withdrawal will be "only 1 per cent of the river" are comparing to flow at the mouth;

critics say you have to calculate the percentage based not on the flow at the mouth but on the allocable volume, i.e. 360, not 1,850. Another controversy concerns just how much water will be withdrawn: 65 cubic metres per second or 127? The project on the books now is still the old model proposed by President Cardoso, built to distribute 127. Pumps, tunnels, canals, infrastructure and maintenance — if you're supposed to take out 65 cubic metres per second, it's all overbuilt. It's like paying ten times the price of the car you really need, just so, once in a lifetime, it can go really fast; sooner or later the driver *will* put his foot down. And when that happens on the São Francisco, all those sheets of calculations will fly out the window and flutter over the *sertão*.

HOW TO LIVE WITH DROUGHT

The drought problem in the northeast is not a lack of water, it's a matter of access to what's already there. With the new diversion project, water will simply be concentrated in the hands of those who already have it. Only 5 per cent of people actually living in the semi-arid area will get any. Most of the water withdrawn from the São Francisco will end up in reservoirs for urban and agricultural use, not for remote backlands areas.

João Suassuna, one of the most persistent critics of the project and himself an engineer, says, *"Vai chover no molhado"* — it's going to rain where it's already wet. It's a wonderful phrase, typical of the Brazilian propensity for flowery speech. It reminds me of Burton's delight in the Portuguese of an elderly gentleman he met during his descent of the São Francisco, who called heat "a temerity of sun" and rich ore a "barbarity of iron." What Suassuna means is that there's already enough water in the northeast, much of it stored in very wet reservoirs. After a hundred years of trying to solve water problems in the northeast, it's hardly surprising that there are plenty of reservoirs already. Three hundred big ones, to be more precise, and around seventy thousand small ones. This adds up to thirty billion cubic metres of water storage, a third bigger than the entire Três Marias reservoir and more than twenty times what the people of the northeast need to reach the UN standard of water requirement. It's enough water to irrigate more than 700 million hectares, but today irrigates only 120.

Critics of the diversion pound incessantly at this single nail: there is already enough water captured in reservoirs; to conquer thirst, all you need to do is connect them, at low cost, to the thirsty people. In Ceará, the enormous Castanhão reservoir is the crown jewel in an extensive network, and it holds half as much as the entire Sobradinho reservoir. Castanhão is the perfect example of poor distribution, or at least a completely undemocratic one. When Castanhão was opened in 2003, after eight years of construction, it filled completely in sixteen days of rain in January, and the spill gates had to be opened. It doesn't need — can't even hold — more water.

But only thirty kilometres away from the huge Castanhão reservoir, in the town of Sitio Quilombo, there is no water supply. New federal settlements, provided after homes were inundated by the new reservoir, have ample treated water and sewage treatment, but the situation in rural areas like Sitio Quilombo is different. Here, the *carro pipa* is still the means of getting water, or the back of a burro or desalinized water at ten reais for twenty litres, if the machine is working. All that's needed to change the situation is a few kilometres of pipe. The small reservoirs are poorly shared too. Networks of smaller reservoirs made by damming a local river or stream have provided water in arid areas all over the world for centuries, and they continue to do so in northeast Brazil. The problem is, 70 per cent of these existing reservoirs are not available to the public.

So it's not surprising that alternatives to diversion of the São Francisco focus on finding better ways to store and deliver rain that falls in the northeast. In dry years, replacing the infamous *carro pipa* with a system of pipelines — in other words, connecting reservoirs with people — is the most commonly advanced solution. When there's rain, cisterns and *açudes* (small reservoirs) are said to be enough; all you need to do is build more of them. Cisterns are a proven technology, and the simplest ones don't even need pumps, just a way of channeling the rainwater that falls on people's roofs. If the poor need water, why not help them get it directly, from the sky?

It costs a lot less to build a million cisterns than to construct canals. The program "A Million Cisterns" managed to build 50,000 in a single year, helping 250,000 people in all kinds of ways, including improved health and employment. By the end of 2007, around 220,000 had been built. The program is funded by the Brazilian government as well as the UN and other agencies; you can even donate through Oxfam. Cisterns and pipelines are examples of "living with drought" rather than combating it. Cistern enthusiasts in Brazil want to follow up with another tank program, this time for supplying small farms, following the demonstrated success in China, where 2.5 million of them have been built.

Watering crops isn't something the cisterns are set up to do, but there are also innovative ways of dispersing irrigation water that reduce waste and evaporation. One is the mandala system, in which a tank of varying size (8,000–30,000 litres), twenty feet across, stores water pumped up either from the closest stream, the kind that runs only when it rains, or from an *açude* or cistern. Around the tank there's an enclosure for the farm animals, and you can even raise fish and ducks inside it for food and for fertilizer. Outside the animal zone, a system of cheap plastic pipes radiates out in concentric zones, each of which is planted with a different crop. It's drip-irrigation using stored rainwater, it creates a small oasis, and it works to feed a family, even provide a little for market. The mandala system is experimental and aimed at organic production and it has support from foundations, government and big companies like Bayer. Its shape is probably not the point — what matters is that you can survive using the water that falls from the sky if you store and distribute it right.

Açudes store much more water than cisterns, but they have a couple of disadvantages: evaporation and access. *Açudes* are open to the sun, and because evaporation concentrates salts, these small reservoirs also add tremendously to the problem of soil salinization. Politically, *açudes* concentrate power as well as water, because they control access to water. It's a method that was brought by the Portuguese and has ensured the extraordinarily long life of the *indústria de seca,* in which drought is used to convince taxpayers to fund water schemes in the northeast. Usually the schemes — canals, reservoirs, dams, wells — end up benefiting the proponents.

WHAT TO DO?

The polemics on either side of the diversion issue are strident. Diversion will be the salvation of the northeast; diversion is just a more sophisticated version of the detested *indústria da seca*. Richard Burton could see the polarization happening way back in 1867, when he stumbled across a proposed diversion of the São Francisco by one Dr. Marcos Antônio de Macedo of Boa Vista. The project would divert water into the Jaguaribe River and thence to the drought-stricken states of Ceará, Paraíba and Rio Grande do Norte. When Burton asked the people of Boa Vista about it they had never heard of Dr Marcos or his canal. Anticipating a century of arguments about the São Francisco, Burton had this to say: "Even were the canal to fail, the strong current of currency generated by the attempt would doubtless bear fruit." Now, 140 years later, proponents are branded as hypnotized by a mirage; the condemnation of History is invoked. Barbara Tuchman's examples of folly and wooden-headedness are everywhere.

For me, the proposed diversion project became one of those sagas that might well never have an end; after reading endless articles and talking to anyone who had an opinion (which was everyone) it had begun to look like a black hole. After a while, the arguments repeated themselves, and the whole issue began to seem less like a battle to be decisively won or lost, and more like an object lesson I couldn't figure out yet. Maybe it was even one of those dark jokes Brazilians were so fond of — except that I didn't know the punchline.

The biggest problem for me was that my experience of the São Francisco was confined to Minas Gerais (the upland state where most of the water comes from) and the wetlands around its mouth. About the northeast, the place where all those 25 (or 63, or 127) cubic metres per second of diverted São Francisco water would end up, I only knew what I had read. And that wasn't good enough.

CHAPTER 14

WATER EVERYWHERE

The Counsellor's country

Two kinds of water storage: cactus garden and cistern, Pernambuco, northeast Brazil.

VISITORS TO BRAZIL'S NORTHEAST SELDOM HEAD INLAND. They come for the beaches and Carnaval, to the cities of Salvador, João Pessoa, Maceió, Recife. The interior is different. To say that it's a state of mind is risking cliché, but surely what we call a state of mind is the result of geography and history and culture. In the northeast these three combine powerfully, as in no other place I've been. The first time I went to the interior I already knew something unusual was about to happen, even allowing for the fact that it was Carnaval season. The unlikely agent was a man called Paul.

He was staying at a *pousada* in the coastal city of João Pessoa, where I was spending a few days at a fisheries conference before heading inland. The annual meeting of the Brazilian Society of Ichthyologists had a surprisingly large body of delegates. It was a good meeting. I took notes on the odd political difficulties of tracking the migratory movements of fish using radio-tags (the locals catch the fish and steal the tags, suspecting that researchers are actually the police), on the effects of introducing alien fish species (Brazil is a leader in this dubious area). I also wrote down the arcane arguments for and against prohibiting biopiracy of ornamental species from the Amazon. The presentations on tracking the migratory fish were especially good: *surubim* averaging twenty-nine kilometres a day through the rapids at Pirapora; the sleeker *dourado* cruising along at more than twice that speed. This was brand-new information, field results obtained against formidable odds and vital for managing what was left of these species.

It was all good stuff, delivered in the second floor of the hangar-sized concrete convention centre in the middle of the city. Outside, fifteen taxis waited for business, their drivers hiking their T-shirts over their guts to keep cool. One was washing his car with a water-bottle and rag; I wasn't sure whether this was a model for water conservation or just a half-hearted way of killing time. Inside, practically everyone was in beach wear, as though they had just wandered up from the Avenida to slip on their identification tags and catch a session on gill parasites. Carnaval finally gained the upper hand when a group of drummers set up in the huge foyer. Posters on pH levels and induced spawning were cleared away and *samba* took over. It was loud; delegates danced; even those trying to maintain a scientific discussion flexed their knees where they stood. Time was suspended. If you were scheduled to deliver a presentation that afternoon you had the choice of talking to an empty room or taking a lengthy break.

The hotel where I was staying had a half-dozen rooms on two floors, a friendly family in charge, hammocks and chairs out front, and a cook who would make any kind of eggs for you in the morning. I first noticed Paul one evening, drinking beer and watching the littoral parade of humanity go by. He was speaking what turned out to be Danish on a cell phone. I envied him that part; the phone I'd rented ("Send or Receive Calls Anywhere in Brazil!") had turned out to be useless outside São Paulo.

Paul was thickset but fit for fifty and obviously proud of it, because I never saw him in a shirt. He appeared to be alone. His Portuguese was sketchy but his English was good, and in a few minutes I pieced together his story. He spent the summers in Brazil, selling Chinese-made costume jewellery door-to-door in what looked like a kind of Tupperware-party arrangement; what passed for summer in Denmark he spent back home. He smoked non-stop and, from time to time, as he described it, "emptied the bar" at the hotel. On Sundays, he crossed the street to join a vicious game of volleyball on the sand, returning to drink steadily with both teams.

"Yah-yah," he said, nodding rapidly, when I told him I was heading into the interior. It was a verbal tic, Paul's "you know" or "um."

"The interior, you gonna love it there. Yah-yah." This surprised me. From what I had learned from all those articles and books, the interior was a blasted plain peopled by scarecrows.

"Pretty dry in there?" I asked.

"Don't know about dry," Paul replied, dragging on his cigarette and shooting the smoke out the corner of his mouth. "Didn't notice. But the people are great. You gonna be treated like a king." He stubbed out his cigarette, stood up and announced he was off to convince Brazilian women to buy his jewellery. I said I hoped he would put on a shirt. Then I took some phone cards to the phone booth across the road to call the Mobil Brazil people again. Half the buttons on the pay phone stuck, and the huge booth cut off any light. I talked to three different people, all of whom suggested a different way to reprogram the phone. And I was almost there too, searching for something called the "tic tac toe button" and holding the receiver against my shoulder, when the phone card ran out. I called it a night.

My progress the next day, away from João Pessoa and inland toward Campina Grande, might not have been regal but it was far from unpleasant. Campina Grande is the biggest inland city in Paraíba, and you have to go through it to get to the more interesting towns of Pombal (where The Counsellor, Antônio Conselheiro, camped out, accumulating pilgrims) and Souza. Campina Grande is slated to get some of the São Francisco water after the diversion. I got lost there for almost an hour, the only landmark being the sprawling women's prison on the outskirts. The prison was a barnacle on a rocky hill, a walled, hopeless place. Campina Grande seemed to be a favoured spot for kidnappings; that night, flipping through the channels in a hotel in Patos, I saw a report that two more people had been grabbed. Kidnapping is a reliable income generator in Brazil, and often involves holding you until your bank card can be emptied. Anyone trying to empty my Royal Bank of Canada card was going to be deeply frustrated; I'd never got the thing to work in Brazil.

Campina Grande is on the edge of the *agreste* zone, a transitional ecosystem that bridges the coastal "forest" (now long-since replaced by sugar cane) and the true badlands of the *caatinga,* a land of boulders and thin soil and fantastically evolved plants that manage to exist in an area with low rainfall and high evaporation. On either side of Campina Grande is a succession of small towns, each announced by a Polícia Militar checkpoint and a series of speed bumps. The police get handled with a handshake, but the bumps are attended by gangs of young men brandishing bags of roasted cashews. The car quickly filled with nuts.

An hour past Campina Grande, the landscape changed again, expanding somehow, as though giant hands had pulled the corners of the earth, stretching it to dry beneath the sun. The strange confinement of the coast — buildings or sugar cane on one side, the unrelieved blue ocean on the other — gave way to a sensation of sailing out of sight of land, except that here, in the *caatinga,* the land itself was the ocean, rolling here, crested with hills and boulders there. But it was far from featureless: cactus stood out against the blue sky like candelabras; when I stopped to pee, the ground at my feet was encrusted with spiky little teapots. How could anyone farm here? Grey rock erupted through the thin soil, splitting the surface like decaying teeth. Much of the un-punctured area was littered with stones that ranged in size from basketballs to trucks.

The stony plains outside the town of Aparecida were flanked with fantastically shaped hills — cones, breasts, shark fins — and goats moved through the streets. I stopped there for a northeast lunch of grilled chicken, spaghetti, rice, beans, salad and what Burton called "the inevitable coffee," which I took with a few hundred cashews. Two little boys cleaned off the shreds of meat I'd left on the chicken bones, waiting politely until I had left.

A week before, when I had driven from Belo Horizonte to Três Marias, the rain had fallen for the entire four-hour trip, grey sheets of it that bounced off the greasy highway and hid the potholes until the last minute. Sometimes the only warning was the behaviour of cars ahead, scattering suddenly like frightened antelope. Truckers saw the holes too late, wrenching their laden rigs, colliding head-on or jackknifing into the shoulder to overturn like monstrous beetles. I saw five major accidents on that leg alone. But the main road heading east into the interior of Paraíba was excellent, recently paved and unclogged. In Paraíba, all was open; the stinking charcoal trucks of Minas were nowhere to be seen because there was no

The "Fish River" waits for rain, Paraíba.

vegetation to feed them, and the sun had burned the landscape clean. Even the vultures had to work fast here, tearing the guts out of fallen cattle before the sun tanned the remains into a leather-clad scarecrow. The living ones moved slowly over the brown stubble, their only other feed the dinner-plate sized leaves of cactus the farmers cultivate in scraggly plots behind their houses.

The little towns appeared and disappeared, announced always by the same silhouettes on the horizon: church, water tower, windmill. Farther on from Aparecida, in an even smaller place called Junco de Serida, the road wound upward to the top of the plateau where the temperature dropped and the vegetation was greener. A small mine was marked by piles of slate and crushed rock lining the road. But what stood out most in this desiccated landscape was, paradoxically, the water.

I had expected a desert, heat-shimmering and forbidding, with perhaps the occasional pond; what I saw instead was one reminder after another that water not only fell here, it was diverted, corralled,

pumped, stored, transported. Most of the small rivers were near-dry:
the sign for the ironically named Rio do Peixe (Fish River) was rid-
dled with bullet holes lit by the fierce sun. But at least some of the
water that ran through them in the rainy season had been captured;
the evidence was everywhere. Gleaming water silos thirty feet high,
with the logo of the state water board. Stacks of six-inch PVC tubing
ready to be connected (or pilfered). Ponds large and small, by the
side of the road and shimmering off in the distance, the small ones
often half-empty, with cattle picking at the dusty bowl. At one of
them I saw a burro waiting while its owner filled scuffed-looking
carboys; at another, a truck rigged with a lozenge-shaped steel tank
was parked at the margin, a black hose submerged and sucking
busily: the *carro pipa*. One big reservoir was a floating garden; a horse
cruised contentedly through it, cool to its shoulders and cropping
on tender shoots. The windmills were mostly of the same design,
metal-vaned and set up near a pond; you could see the piping
snaking uphill to the storage tank perched on a farmhouse roof. The
critics of the diversion seemed to be right on this point: there *was*
water out here.

I got good at spotting water tanks. There were many shapes and
sizes, made of concrete or brightly coloured fibreglass. In the small
towns, houses and businesses had one on the roof; in larger centres
you could spot the big municipal tank from ten minutes out, looming like a pillbox on stilts, or concrete and square and perched
fifteen metres up like a windowless bunker.

In the town of Santa Luzia, the municipal tank was older, and
quite spectacular. The town is reached after winding up through a
valley. I could hear the hooves of two horses picking their way along
a dry riverbed that had become a road, bouldered and sandy. But farther ahead the small dams appeared, then the reservoirs; Santa Luzia
itself seemed to have captured most of the water behind a municipal
barrier that formed a pleasant lake beside the town. At the highest
point of this pretty place there was a stone water tank, round, with
gently sloping stone sides and a flat platform for a lid that looked as
though you could land a helicopter on it. A massive outlet pipe dove
beneath the cobbled pavement, presumably fanning out to the neat
houses where families were gathered in gardens. I sat down and ate

Water storage in Santa Luzia, Paraíba.

some more cashews, watching the vigorous game of billiards going on in one of them.

This was confusing. I tried to reconcile everything I had heard and read with the evidence of my own eyes. The northeast was supposed to be dry, and its rivers certainly seemed to be, exactly as promised, but it was just as obvious that here existed a culture of water storage, water transport, water distribution. The question was, did it all work? And if so, who got the water? To people in developed countries, especially a water-rich one like Canada, basic questions about water simply never arise. In Victoria, where I live, a shortage is when residents are forced to suspend watering their lawns for two days a week in high summer. The water still comes from a tap and the tap is connected to pipes, which coalesce like capillaries running backward to the heart, all the way to the reservoir in the hills. That's it. So when confronted with ponds and windmills and storage tanks large and small dotting the countryside and perched on roofs (I saw a cheerfully painted pair of them on the roof of one of Brazil's love hotels, done up in blue waves and a brilliant yellow sun), when tanker trucks nose through towns and people lead horse-drawn carts piled with carboys through the streets, you have to ask a whole lot more questions. Who uses this water (people? animals? crops? fish?); where does it come from (river? pond? reservoir? well?); how much does it cost; who do you pay; and in what currency (money? votes?). And is it safe to drink?

Obviously there *were* water problems out here. In Patos, a big town near a ceramic works where I spent one night, I stayed at the Hotel JK. Despite supposedly being the largest in the city, the hotel was difficult to find. *"Zhota kah!"* people kept telling me when I asked for directions. "You can't miss it!" followed by the usual spectacular body language of Brazilian directions, like a sailor semaphoring on the heaving deck of an aircraft carrier. Off I would go, always to the same place, until I finally remembered the Portuguese alphabet: JK, *zhota kah,* Juscelino Kubitschek, President of the Republic and builder of Brasilia. Owner of the most recognizable initials in the country. There they were in fact, three metres high on the side of the building. I felt pretty stupid.

JK *was* the biggest hotel, but when I stepped into the shower in the morning it reeked of sewage, as though the tap were connected somewhere unmentionable in the bowels of the building. Eventually the taps ran clean, when the water stored on the roof had flushed out the pipes. The bottled stuff was no better, smelling strongly of kerosene. "It happens," shrugged the man at the desk.

But there *was* water. I wrote in my notebook, "System for storing and transporting water. Do you just need to build more of the same? Or a better system? How will SF diversion help?" Driving back toward the coast, I thought about these questions, and I began to feel like a man returning from a puzzling blind date.

"Don't get your hopes up," someone might have advised me, setting out. "She's, well, not ugly exactly, but she isn't beautiful either. More like plain."

"Okay," I would have replied. "It's the personality that matters, right?"

Well, the northeast seemed to have personality in abundance, and if she wasn't beautiful she was something even better: she was striking. And she was mysterious, questions upon questions. Here was the response to the São Francisco diversion, right in front of my eyes. I just had to decode it.

THREE CARNAVALS

I arrived back in João Pessoa two nights before Carnaval. And just like that, as so often happens in Brazil, the sublime took a back seat to the ridiculous. Paul was already at the bar, and he'd found some assistance in emptying it this time, an Israeli woman half his age who was travelling the world with her surfboard. She was leaving for Fortaleza tomorrow, up the coast. She and Paul seemed made for each other.

The chairs in front of the *pousada* were occupied by the landlady, a vivacious middle-aged woman with skinned-back hair and a dazzling grin, and a young man dressed in a bikini. I sat down and offered each a cashew. The man's bikini top consisted of two coconut halves linked with gold chain. I tried not to look at the bottom. He uncrossed long shaven legs and leaned forward, chewing. He grasped my wrist.

"You'll dance with me, won't you?" he said. "Somebody's going to dance with me." He wriggled in his chair. "Oh, *many.*"

I took my revelations about water management in the northeast and shoved them temporarily into a mental corner. If you're in Brazil during Carnaval it's hard to get much done; if you're back home, your Brazilian correspondents tend not to correspond. Here in João Pessoa, the entire Avenida Cabo Branco, the five-kilometre beachfront promenade that passed inches from our *pousada's* front door, was readying for Virgins' Night, a pre-Carnaval tradition in many parts of Brazil but especially strong here. During Carnaval itself, still a few days off, the Virgins would be just another *bloco* or organized group that paraded along with all the other *blocos,* dancing and drinking and absorbing kidney-liquefying levels of sound from house-sized, truck-mounted loudspeakers inching along behind. Tonight, though, they would have João Pessoa to themselves: a captive audience, me included.

Resistance was futile; everything was slipping toward Carnaval. I decided to have another go at reviving my rented cell phone before the entire country went down. I bought another handful of phone cards and crossed the street again to the *orelhão*. It was considerably noisier now, but I was determined to make my cell phone work. An entirely new representative in São Paulo admitted it was all his fault and he would call me back in ten minutes with the solution. I decided to wait and see. Maybe Carnaval already had Mobil Brasil in its grip.

By nine o'clock, nobody from Mobil Brasil had called back. I ducked back into the *orelhão* determined to keep negotiations alive; now it was deafening. The phone cards bent going into the slot. When I got someone on the line (of course, a different person) I had to shout. The guy shouted back; maybe he was in the middle of a party too.

"Okay, *Senhor* Brian, no problem! We gave you the wrong phone! Where are you staying?" I began to answer and the mouthpiece on the pay phone broke spontaneously in half and clattered to the pavement, or at least I assume it clattered, because I couldn't hear anything. I bellowed the address into the guts of the phone and hung up. The receiver cradle fell off. A kid went by on a bicycle, talking on *his* cell phone.

Virgins' Night was getting up a head of steam by about ten o'clock. The general idea was to promenade up and down the Avenida for an hour or two, sampling street food, dancing and drinking and comparing notes on the other participants, eventually collecting at the bandstand a kilometre or so down the road from the *pousada*. Seen from high above, we must all have looked like ants in a nature documentary, scurrying back and forth along a scent line, feelers waving, returning again and again to the point where the smell was strongest until finally the whole swarm was massed there, vibrating busily.

The virgins' costumes varied depending, I guessed, on the degree of commitment to the concept of dressing up in women's clothes. Many of the guys had obviously just borrowed clothes from their girlfriends or sisters (probably, in the case of the beefier ones, from mothers too), and were happy enough to stroll down the Avenida in too-tight miniskirts and stuffed brassieres, with fright wigs balanced over foolish grins. Hands were still for the masculine purpose of carrying beer, and feet were still for walking in the normal way, in

Some virgins at Virgins' Night, João Pessoa, Brazil.

tennis shoes or, at the very most, sandals for that open-toed look. These guys, the dabblers, travelled in groups, blowing kisses and drinking steadily. Their breasts tended not to be level. One wore only a corset and had got himself pregnant, but he still retained his horn-rimmed glasses over the glittery eyeshadow. I have a treasured photograph of three of them flipping their skirts up, forming an instant, shaky pyramid and mooning for the camera: six hairy, sweating, thong-split buttocks.

But the wonky-boobed were amateurs. The real virgins didn't carry beers, they carried handbags, and they travelled alone or at the most in pairs. They dressed like women, right down to the spike heels, and they walked like women. Some invited a response, like the guy gyrating in a miniskirt, tongue flicking, or the one in an electric blue sheath who suddenly dropped to the pavement and spun on one palm like a stripper, instantly clearing his own little stage down there with the spilled beer and the cigarettes and the fake dog poop. The best of them were above all that, parading haughtily like unusually muscular fashion models. My favourites could have been twins, with ironed hair, identical in stars and stripes miniskirts and halter tops above stiletto heels.

The evening wore on; nobody paid any attention to me even though I seemed to be the only *gringo* in João Pessoa. Deafening music from roaming sound-bicycles blended and blurred. Some of the vendors sold grilled things on sticks: chicken, shrimp, fried cheese of course; others offered tapioca wraps brimming with red syrup, fiery Bahian pastries, boiled corn, chocolate-dipped fruits and so many coconuts the margins of the stroll began to accumulate piles of green husks. A lone Asian vendor stood grimly beside his noodle stand, ignored. Watchers looked down from their apartments and hotel rooms and the Atlantic lapped invisibly at the sand.

In the morning, it was all gone. At six AM the Avenida was spotless, and the only people around were joggers. I had a flight to catch. As I was loading up the car the desk clerk came running out and handed me a courier package containing a small miracle: a new cell phone, overnight. I switched it on and the signal metre came up strong, as if to say, what exactly was *Senhor* Brian's problem?

In the departure lounge at the airport, I began to review my notes about water. I knew now what I had suspected ever since the São Francisco diversion had hit the headlines: selling the project as "water for those who have none" was beginning to look like just that — a sales job. The places I had visited over the past few days, before Virgins' Night had pushed them to the back of my mind, received more annual rainfall than Paris, more than Winnipeg, more in some years, for God's sake, than my own home town on the "wet coast" of Canada. So it couldn't be just the lack of rain. I had already begun to plan my next trip to the northeast when a tired-looking young man sat down beside me. He wore low-cut jeans beneath a flat, shaven belly, a white sleeveless blouse and a Gucci bag. Despite the early hour, he had still managed to do his makeup and earrings.

"*Bom dia,*" I said, and he managed a wan smile. I gave him my last bag of roasted cashews.

Two days later I was back in Pirapora, where I hoped to see a "real, rural" Carnaval. I didn't. That was the year the town decided to experiment with something snazzier, more commercial, and when I wandered down from the Hotel Cariris at nine o'clock the centre of town was behind a ten-foot fence. Security guards with metal detectors were frisking the young men passing through the only opening (they left the women alone). I hung around, sampling sticks of chicken cooked not over the coals of João Pessoa but on griddles swimming in smoking oil. In one kiosk, a waiter chopped ice cupped in one hand, using a bottle opener.

On the main stage, the one where fabulous bands were rumoured to appear extremely soon, a man shouted innumerable announcements, thanking mayors, sponsors, the Department of Health. Everyone over the age of sixteen appeared to be overweight. Girls wandered past with cell phones tucked between brown breasts. A scruffy ten-year-old worked methodically through the crowd picking up bottles while more kids played a brilliant game of soccer using an empty Coke bottle.

Finally the *blocos* began to emerge from somewhere else in the city, the progression of an extremely loud snail. The first one was Forro Velho — more men dressed as women! The water in my bottle vibrated from the sound coming off the speakers. Hips waggled, people sweated, kids shot strings of instant Styrofoam over me from aerosol cans, my ears hurt. By midnight, no bands had appeared, and the man shouting announcements appeared to be thanking his mother-in-law.

The next day I drove back to Belo Horizonte. Carnaval in Belo was much simpler: everybody left town, for the real celebrations in Ouro Preto, Diamantina and, I found out too late, Pirapora. In Belo, all was tranquil. Carnaval was the televised version from Rio, live twenty-four hours a day. I relaxed with Hugo's extended family and ate ice cream. The women offered commentary on the quality of the silicone breasts. It was all very civilized. And that year, as a special treat to the masses, the Ministry of National Integration, the arm of government promoting the diversion project, sponsored its own float: a celebration of the São Francisco River.

CHAPTER 15

DANCING WITH THE BISHOP

When politics and engineering meet
a Higher Power

In Brazil, religion is never far away: fuel truck on a perfect highway in Paraíba.

I IMAGINE A BALL. Something baroque and formal, with a beginning and an inevitable end, but divided into movements, each with its own character. It happens in an eighteenth-century ballroom, gilded and vast. The court musicians adjust their hand-copied scores of Lully or Rameau, the rustling of crinolines and cummerbunds dies to a whisper, the leader of the orchestra raises his bow. Then the downbeat, and the first collective step: the *Prelude* begins. The dancers advance, retreat and circle, passing from hand to hand, never departing from form. Kings and queens, courtiers and grandees, usurpers and illegitimates — if you were a child peeking through balustrades high above they would all look the same. All smiling, all twirling.

That's how I see the colonial history of the São Francisco River: a dance across the centuries, the partners willing at times, barely speaking at others, but dancing nonetheless. The first steps were taken in 1501, when the mouth of the river was discovered by the Italian Americo Vespucio and named for the saint on whose day it was first seen: St. Francis, protector of birds, plants and animals, patron saint of ecology. Two hundred years later, the northeast suffered a drought so disastrous the Portuguese Crown considered the idea of building reservoirs to conserve water.

In 1824, the river got a closer look. At the request of the Portuguese Emperor Dom Pedro II, a French explorer and naturalist was engaged to survey the flora and fauna of Brazil from Jequitinhonha to Rio de la Plata. His name was August François

César Prouvençal de Saint-Hilaire. One chapter of his resulting book was a visit to the source of the São Francisco River. Saint-Hilaire returned to Europe with a vast collection of plants, birds, insects, mammals, reptiles, fishes and minerals. An original copy of his *Brazilian Flora* can now be had for $US 28,000.

But the problems of the northeast continued. In 1852, Emperor Dom Pedro II declared he would sell the crown jewels to end the suffering of his subjects in the northeast. He ordered the first studies of the possibility of withdrawing water from the São Francisco River and transporting it to regions "castigated" by drought. The engineer Henrique Halfeld selected the area around Cabrobó as the best place to withdraw water, a choice that remains to this day. But nothing was done, and the Emperor kept his jewels. However, four years later, the Brazilian Institute for Geography and History produced the first formal proposal for diversion of the São Francisco.

1867: Sir Richard Francis Burton travelled down the São Francisco by raft, writing furiously. Of the river and its future, he said, "The valley and the high dry Geraes which limit it on both sides contain all the elements of prosperity required by an empire . . . they are emphatically the Lands of Promise." Ten years later, an historic drought killed 1,500,000 *nordestinos*. The first *açudes* were finally built in the northeast — a hundred and fifty years after the idea was raised. In 1906, the writer Euclides da Cunha, chronicler of the bloody putdown of the revolt at Canudos, resurrected the idea of diverting the river. Five years later, the first Brazilian dam was built at Delmiro Gouveia on the lower São Francisco River (it's still there, stuck like a limpet in the chasm near Paulo Afonso, a tourist attraction now).

It's a fine *Prelude,* full of the promise of great things. In four hundred years the dancers have created a new country, its geographic focus a river that's the only perennial source of water in a land of drought. Solutions to the problem of drought have been proposed, muscles have been flexed. High above, on the landing, the child rubs its eyes. The dancers pause while the musicians refresh themselves, casting appraising glances around the ballroom, noting adversaries and allies. The musicians re-tune their instruments for the next movement, the *Courante.* Then the dancers twirl on.

In 1985, the first modern proposal for diverting the São Francisco emerged. The now-defunct National Department of Public Works and Sanitation (DNOCS) proposed the withdrawal of an absurd 25 per cent of the river's flow at Cabrobó, between the Itaparica and Sobradinho dams. The project went nowhere, but it wasn't forgotten. In 1994, Luiz Inacio Lula da Silva ("Lula") took his second run at the Brazilian presidency. His opponent, Fernando Henrique Cardoso, defended the idea of diverting water from the São Francisco to the northeast. The bearded Lula, champion of the workers and himself from the northeast, promised, if elected, to study the alternatives. He lost. A year later, President Cardoso collected a symbolic sample of water from the Serra da Canastra, the source of the river, and declared, "The river is generous and doesn't have to dry out in order for the northeast to take a little." Technical studies of various diversion schemes began to appear in academic journals, and a new project took shape. In 2000, the Brazilian government heard the ghost of Richard Burton whispering from the pages of his book: "It is only a question of time when the Brazil will follow the example of the United States." The Minister of National Integration, Fernando Bezerra, accompanied by members of the Chamber of Deputies, made the fateful study tour to Colorado, organized by the U.S. Bureau of Reclamation. When they got home the Ministry produced the *Projeto do Transposição do Rio São Francisco para o Nordeste Setentrional* (PTSF), shortened to the "São Francisco Project."

Now the pace of the dance has quickened. In just fifteen years, a plan has emerged — one worthy of royalty, or at least of those with kingly intentions. The dancers are flushed; the air crackles. It's as though the planned order of the music has changed: what should have been a stately *Sarabande* will be another kind of dance now, perhaps a *tarantella*. The dancers have split into two camps. In 2001, Senator Paulo Souto of Bahia (a donor state) released a detailed document in the Brazilian Senate: "Transposition of the São Francisco: An Unsupportable Project." His report detailed fundamental flaws including misconceptions about available flow to be tapped, jurisdictional chaos, the high cost of construction and evaporation losses, and the many alternatives that place less emphasis on irrigation. Above all, the report attacked the truthfulness of promoting "water for the people" when 70 per cent of the redirected flow has already been assigned to irrigation. In the lively debate on his report, Senator Suassuna of Paraíba (a receiver state) asserted his state's "biblical right" to the water. In the same year, the World Bank released its damning report on the diversion project.

But Lula had finally won the presidency, on his third try, on promises of "sustainable development." The PT ("workers'") party was in power at last. There was no mention of studying any alternatives to diversion, and by 2004 the drought of 2001 seemed to be forgotten. The diversion picked up steam. President Lula cast the project in historic terms. Not to build it would be an "historic error." It would bring about a "revolution" in the northeast. He announced the first phase, an investment of $2 billion reais. The São Francisco Basin Committee went on the road to five major municipalities along the river to hold public consultations. Four thousand people showed up; all voted "no" to diversion. In Salvador, the Committee went head-to-head with the national Council of Water Resources. In Belo Horizonte, the diversion project was overwhelmingly rejected. A Manifesto to the Country written by eight specialists in engineering, geography and sociology from both donor and receiver states was released in Brasilia. Highly critical of the diversion, it was signed by 210 associations, institutes and NGOs. Front-page articles on the diversion appeared in major Brazilian newspapers.

By 2005, resistance and rhetoric were escalating. Lula insisted on the project, saying that only 1 per cent of the river's water would be withdrawn. The Ministry of National Integration compared Lula to the great Juscelino Kubitschek. Minister Ciro Gomes went on the stump for the project, assuring radio and TV audiences that the water to be withdrawn from the river would otherwise simply flow out to sea and today "is used for absolutely nothing, not generating any benefit for anyone." Public meetings on the environmental licencing of the project in donor states collapsed. In Belo Horizonte, Salvador, Aracajú and Maceió the meetings were disrupted or boycotted; the one in Belo Horizonte featured whistles, music, placards and the symbolic burial of Ciro Gomes and the Minister of the Environment, Marina Silva. This "legal process" continued and the

Ministry of Integration opened bidding for construction contracts. In late 2005, people from eighty-nine popular organizations, NGOS and pastoral commissions met for two days in Bom Jesus de Lapa, Bahia, the symbolic "midpoint" of the river. They issued a manifesto against the diversion and the privatization of water.

Now the dancers have reached the stage when excitement just balances exhaustion. The dance steps are still familiar, but they are coming faster. There are stumbles, and small collisions, and tense apologies over a retreating shoulder. The new King and his courtiers watch intently, looking for weaknesses.

And then it happens. The music falters and disintegrates: a note dropped here, a phrase there, then one player after another lifts his bow. They all look in the same direction, mouths open, bows suspended over the strings. On the landing, where the child was hiding, now stands a figure in a simple brown robe that reaches the floor. The King groans, the sound of a man against whom the scales have just been drastically tipped. God has just entered the building.

Twelve years after his famous fact-finding walk down the length of the river, the friar is on the balcony looking down at the dancers. Except that now he is *Dom* Cappio, Bishop of the town of Barra in Bahia, and he has decided that the dancing at the expense of his beloved river had gone on long enough.

THE NEXT LEVEL

In late September of 2005, Dom Cappio announced a hunger strike in protest against the diversion. In an impassioned letter to President Lula, he begged for attention to the real problems of the poor of Brazil's northeast, and called for a policy of living with the semi-arid surroundings. The hunger strike, he was at pains to point out, was a protest against the "insane" plan for water transfer, and not against President Lula. His letter concluded, "My life is in your hands." On the same day, the National Water Agency granted its Certificate of Sustainability to the project. Still on the same day, the artist Bene Fonteles, of the Movement of Artists for Nature and one of the four participants in the historic walk from the source of the river to its mouth in 1993, published an open letter "to the river" in which he compared its sufferings to those of Christ on the cross, and denounced the diversion as bleeding the lifeblood of the river to serve the needs of the market. Still on the same day, Pedro Brito, charged with general coordination of the transposition, proclaimed that the project was "technically perfect, socially just and environmentally sustainable."

So the battle lines were finally drawn, and on one side was the Church, or at least one of its representatives. This should be no surprise, because religious activism is part of the Brazilian cultural fabric. In the mid-1980s, land reform — the redistribution of unused land to sharecroppers that was at the heart of the *Movimento Sem Terra* — resulted in a long list of killings before the death of a thirty-three-year-old priest, Father Josimo Tavares, forced the government to enact legislation for redistribution of land (he was the fourth to die in two years). Immediately after Dom Cappio announced his fast, the Pastoral Commission and the National Confederation of Bishops of Brazil issued statements of support. The bishop moved to Cabrobó, a town on the São Francisco in Pernambuco and the site where water will first be withdrawn from the river; visitors and supporters began to arrive. The hunger strike was described, in language Richard Burton would have appreciated, as "a toad in the Government's throat."

A few days later, Dom Cappio received an emissary from President Lula. The Bishop responded by confirming his decision to starve himself for the river. He released an open letter "to the people of the northeast," warning them the diversion would not solve their problems, but only extend the reach of large-scale irrigated agriculture. A Web site appeared (www.lifeforalife.com) in support of Dom Cappio. The site provided commentary, articles, pictures and links for donations. In one picture, Dom Cappio appeared to bless a sandy-looking outline of the Brazilian northeast, upon which was superimposed a puddle of water in the shape of a head. On the first two days, more than three hundred electronic messages of support were published on the site. The Governor of Bahia arrived and declared his support.

Religion, history and politics were now powerfully joined. Over the next few days, Dom Cappio continued to provide interviews, reiterating the need for other solutions like cisterns and small reservoirs. The pilgrimage to Cabrobó began with buses arriving from a dozen cities and towns including Salvador on the coast and Petrolina on the massive Sobradinho reservoir. October 4 was a day of powerful convergences, historical, religious and mundane: it was the day of Saint Francis, after whom the river was named; it was Ecology Day in Brazil; and it was also Dom Cappio's fifty-ninth birthday. Acts of solidarity included blessings, orations, speeches from ordinary people of the northeast, and a celebration of the Eucharist by Dom Cappio. The Bishops of Sergipe, Propria, Aracajú and Estancia organized a demonstration on the bridge that crosses the river and links it to Alagoas. A march in solidarity with Dom Cappio took place in Salvador on the same afternoon, terminating in the Church of St. Francis. Later in the day, Dom Cappio performed Mass for four thousand pilgrims, all of whom had fasted.

For me, sitting at my desk in Victoria, the e-mails coming in every few hours were further evidence that the whole controversy was much more than the sum of its arguments. Whether the diversion got built or not seemed to matter less than the impassioned waving of dirty laundry. Religion, land ownership, destiny — with these kinds of overtones, how could anybody make decisions? And how long would any decision last? I dithered about going back to the northeast: grabbing a flight to Recife, renting a car, driving all night to Cabrobó. In the end I picked up the phone and called Barbara Johnsen in Três Marias. She was in tears.

"All the people on the river are crying today. You Canadians cannot understand." She described Dom Cappio as "my best friend." "He is my compass. When life was hard for me, I wrote to him, and we worked together too, for the river." She said she wanted to embrace me so that our compasses were aligned.

The hunger strike lasted eleven days before a deal was made between Dom Cappio, an emissary from President Lula and the Vatican's Ambassador to Brazil. The details were vague; the government pledged talks on "other possibilities" before starting the project. Government also apparently agreed to speed up the release of funds for revitalization. Dom Cappio said he would stop eating again if the deal were broken. The Vatican's ambassador reminded him that taking one's life is against Catholic teachings: "The Church is faith, morals and discipline. All three must be respected." Lula stuck with the rhetoric of "water for the poor," reminding the Bishop of his own boyhood in the northeast. Cabrobó returned to normal. The pilgrims went home; the chapel emptied. It began to look as though discussion on the diversion, which began in the time of Emperor Dom Pedro, would last another hundred and fifty years.

The first fruit of the uneasy agreement in Cabrobó was a meeting of NGOs, social organizations and experts in mid-December in Brasilia, organized by Dom Cappio. Its main aims were to come up with a strategy for guaranteeing water for poor people in the semi-arid northeast, and to ensure that water actually translated into development. A report to the *Folha de São Paulo* from a participant described the two-day event as "emotional" and the "beginning of the national debate on the project." (My heart sank; this was just the beginning?) Representatives of the group met with President Lula afterward, forcing him to keep four hundred delegates at a ceremony on Brazil's Millennium Development Goals cooling their heels. Dom Cappio's group presented their proposals for "living with drought": water reform, democratization of access to the rainwater already stored in reservoirs in the northeast, agrarian reform to help families gain title to land and become self-sufficient, official policies on water-use technologies, including storage of rainwater in cisterns. Lula's response was, "Convince me."

By 2006, I had begun to form the notion that the diversion project would be debated forever. Nothing much happened in January and February (when Carnaval takes over), and most of the rest of the year was taken up by campaigning for national elections in the fall. Various civil actions, brought by donor states and aimed at the environmental licence from IBAMA, kept the project on hold; each one had to be fought and eventually overturned by the Supreme Court. The Army camped out in Cabrobó, waiting for the green light to begin construction on the first canals. But it certainly didn't appear that government intended to shelve the project.

And so it all happened again. In late November 2007, after a construction permit for the first phase of the project was issued, Dom

Cappio stopped eating for a second time. This time he held out for twenty-four days, and emotional response inside Brazil was even stronger, with occupations of government offices in Brasilia and Bahia. But this time it wasn't a deal with the President that ended the fast; it was the Bishop's own family and supporters, who rushed him to hospital when he lost consciousness. Once again, President Lula held firm. As if to drive the loss home, the Supreme Court ruled construction could legally proceed.

I looked back on the events of the previous year. Hunger strikes, helicopters, emissaries from the Palace and negotiations in a chapel by the river: what was all that *about?* The real effect of Dom Cappio's strike, I decided, was to wake Brazilians up. The diversion project had already been laid open, dissected, examined minutely and excoriated for at least five years, but always within a closed circle of NGOs and academics. Their output was like the flood of reform-oriented material — "literature, sermons, pamphlets, songs and political assemblies" — that assailed the equally deaf Renaissance popes. Anyone at all familiar with the issue knew its critics were beginning to repeat themselves, spinning angrily like a fly upside down on the windowsill. With adversaries like this, government was still able to maintain a position of mild annoyance while getting on with the program. When the buzzing got too loud, as it did when public hearings were disrupted by protesters, government could still declare the meetings to have been held — technically at least — and get away with it.

Dom Cappio and his hunger strikes have changed all that. The annoying buzzing became irrelevant because the window itself disappeared, obliterated with a dazzling light that instantly whited out corporeal things like flies and windowsills and canals and flow rates. When the Bishop started living on nothing but water from the São Francisco, he invited faith into the debate, and faith is something even the Ministry of National Integration was reluctant to tangle with. "The last thing I need," said Lula, in by far the most sensible pronouncement to emanate from government, "is the body of a bishop on my back." I imagine him flicking that suddenly irrelevant fly out the window. He leans heavily on his knuckles, rotund and pouchy, staring out over the futuristic Niemayer landscape of

Brasilia, as though he can see all the way across the *planalto* to the baking back-country of the northeast. Maybe Ciro Gomes, vulpine and goateed, joins him and they stare silently together, the rotund president and his slender minister, thinking about all that investment, all those images of order and progress, all those votes.

From the moment the first hunger strike began, the rules changed. Making the debate a religious one brought the media, first pelting full tilt to Cabrobó where busloads of the faithful were converging, then scrambling to contact every expert, every lonely buzzing fly out there, so that for three weeks in October of 2005 the evening news was never without a live report or a round table of experts debating the project. E-mail traffic exploded. An agricultural worker in Pernambuco told me, "I couldn't believe it. I turned on the local radio one day and there was a phone-in about the project. And these were people who'd never even heard of the thing a week ago!"

Pilgrims streaming into a tiny town where a bishop had put his life on the line — this was what the networks wanted to see. It didn't do to look too closely at the pilgrims themselves, but when you did, the new rules of the game became even clearer. A farmer brought an image of St. Francis, saying, "I don't know much about this business of taking water out of the river, but Dom Luíz represents the power of the Holy Spirit, and everyone who follows him will be blessed." And those who don't, he might have added, will be the other thing. Rubbing it in for Lula even more, another pilgrim declared, "God is giving us the strength to win this battle. The bishop is being crucified for us."

But the President had, after all, prevailed. As this book went to press, the São Francisco diversion project continues to shuffle forward, encumbered from time to time by delays for court challenges at various levels of government. Those challenges may in fact be the project's Achilles heel, because courts are not untouched by popular feeling, especially when it's the kind generated by the actions of a man of God. The ironies — the pesky bishop, the slap on the wrist from Rome — make me reach for my copy of *The March of Folly.*

SHORT MEMORIES

The first of Barbara Tuchman's three historical examples of folly is the strange tale of the six Renaissance popes, and how their excesses provoked the rise of the Protestant church and the eventual Sack of Rome in 1527. This was not long after another Italian, Americo Vespucio, discovered the São Francisco River, a time when, as Tuchman dryly put it, "the efflorescence in culture reflected no comparable surge in human behaviour but rather an astonishing debasement." In other words, a period not only of great art and literature — Leonardo, Michelangelo, Titian, Machiavelli — but also of those in power relentlessly pursuing the spoils of office, and indulging themselves to the limit. Tuchman called the church during that period "the supreme pork barrel."

As Vicars of Christ, Tuchman writes, the popes "made their office a mockery and the cradle of Luther." Or in the words of Leo X, "God has given us the Papacy — let us enjoy it." Which of them had been greediest? I thumbed through the popes until I reached the third of them: Rodrigo Borgia, Pope Alexander VI (1492-1503). His reign coincided almost perfectly with the European discovery of Brazil, and for sheer depravity and debasement of the holy office it was unparalleled. Borgia fathered seven children and made sure that his mistresses had husbands; cuckolding them seemed to add to his pleasure. One of his sons, Cesare, laid on an infamous banquet in the Vatican, over which his father presided: fifty naked courtesans crawled through chestnuts scattered between candelabra on the floor, then coupled with the guests. Prizes were given for "those who could perform the act most often."

I imagined the scene: the stone floors, the hot chestnuts rolling around like golf balls, the skinned knees. Soon after, mares and stallions were herded into a Vatican courtyard in front of the Pope and his daughter Lucrezia, who "watched with loud laughter and much pleasure" as the horses got it on too. When Borgia finally died, he was said to have won the papacy through a pact with the Devil; ever since, his rule has been decribed by historians not in terms of religious observance but political wars, manoeuvres and continuous violence.

Goings-on like these annoyed a lot of people. The folly of the popes wasn't so much in pursuing counter-productive policies; it was their virtually blanket refusal to listen to the rumblings of discontent that led inevitably to the Sack of Rome. Obsession with personal gain was the key factor, so that voices like that of Savonarola, the most charismatic of the papacy's critics, were largely ignored. In Borgia's defence, Tuchman points out that prophets "filled with the voice of God are not easily silenced," something that made me think at once of Lula's troubles with Dom Luíz Cappio. Savonarola was hanged and burned (the latter presumably for good measure) in 1498.

Of course, President Lula is not the Borgia Pope, and Dom Cappio is not even remotely like Savonarola. Nevertheless, the human mind (my human mind, at any rate) often makes its most useful leaps by way of parallels and analogies; that was why Tuchman's book proved so provocative and has so many fans. The ability of the Renaissance popes to ignore a deafening chorus of complaint and go about their business of looting the holy office seemed perfectly reasonable to compare to the insistence of the Brazilian government on diverting a river in the face of a storm of reasoned criticism.

The saga of the São Francisco had everything: history, politics, chicanery, God. Everything, that is, but an ending. When I got the chance to return to the northeast and take another look at the country Sir Richard Burton called "a fracas of Nature," I jumped at it. Not because I wanted to collect more polemics on water — I already had stacks of those, and the e-mails from critics of the diversion kept coming in at the rate of five or six a day. This time, I wanted to look at the alternatives. And for that, I needed to find one of those special people who unlock the secrets of a strange land: an expatriate.

CHAPTER 16

SOMETHING
CONCRETE

Death and rebirth in the Brazilian backlands

Old ideas are like fine wine: the beginnings of a cistern near Brejo da Madre de Deus, Pernambuco.

IF YOU WANT TO UNDERSTAND AN ISSUE IN ANOTHER CULTURE, you must not only learn the language; you have to live it. Not just for a three-week visit; not even for a few months. It has to be long-term, at least a year. In other words, you have to become an expatriate.

Most people don't, of course, but you can still jump-start your understanding by talking to one of these unusual people, assimilating as rapidly as possible the experiences of someone who has lived, not just visited, in another place — but is still a product of your own culture. Spending time with such a person is like getting into a glass-bottomed boat: the expat is a magical lens, thrust suddenly into an unseen world. Such unusualness is priceless but it's not unique; there are thousands of expats who have gone native, all over the world. Andrew the field botanist in Thailand, with his sweats and shakes and whiff of soiled bedsheets, was one. In Brazil I'd already met several, including a tiny energetic Englishman who seemed to know everything about fish feed, and a slovenly Irishman, six and a half feet tall with a ring in one ear, whose knowledge of migratory fishes was encyclopedic and whose last posting was in Tristan da Cunha, in the exact centre of the Atlantic.

You can even find expatriates in your own country. When I visited Ahousaht, a remote Indian village off the coast of British Columbia, I felt as though I had landed on another planet; the surroundings were beautiful, the community devastated. It looked as though a bomb had gone off: shattered windows patched with blue tarpaulins, fishing boats rotting in the mud, rusting appliances and metal chairs upside down in front yards. A rabble of dogs, humping and scratching, grocery-laden all-terrain vehicles snarling past, families pushing kids in strollers and the sounds of a soccer game from a hidden field. I spoke the language but even though the tourist haven of Tofino was only an hour away by boat I was out of place: "White men can't dance," said one bumper sticker on an unlicensed pickup flying a rebel flag. Most of the houses had satellite dishes on the roof, next to the remains of Christmas lights. The shallows were littered with car batteries, rusting differentials, bicycles with eelgrass filtering up through the spokes. The boatyard was abandoned, with a faded sign that said, "Welcome to Ahousaht."

All of this I saw on a twenty-minute walk, just a short turn through the village before sitting down to dinner. It was pretty confusing. I returned to my lodgings, a bed-and-breakfast six metres from a moonlit sandy beach littered with flattened pop bottles and faded potato chip bags. Inside, two white guys in dirty socks lay on a couch, watching a movie. A young white woman sat silently at the other end of the breakfast table, taking notes on a laptop. She was doing her thesis on First Nations. And in the room across from mine, an expat: Wayne.

I was there to advise on clam farming, something a lot of Indian Bands were getting into. Wayne's job was to build houses on the reserve. He was doing this alone, occasionally visiting his family south in Nanaimo. His room was neat, with a half-empty bottle of whisky the only thing on the dresser. He had steely short hair and seemed in some other universe, with a transfixing stare. He had been in Ahousaht for six years.

"When are you leaving?" he asked me. I told him: two days. "You're lucky," he said. His voice was flat.

"What do you do here at night?" I asked.

"Wait until morning."

Wayne the catatonic contractor, Andrew the botanist, and all the others have something in common, something fundamental. I'm not sure what it is, even though I have met many like them — perhaps nothing more than a willingness to be where they are, to sever ties in a way that I never could. They don't appear to be mining their new lives for experiences to take back home; none that I know of has ever thought of writing about all the amazing things that have happened to them. The writers are too obviously what they are, with their notes and their questions, their panic to "understand," their deadlines; people like Wayne and the others just live it. In ten minutes, I learned more about Ahousaht from him than I could have picked up in ten days on my own. Now I was returning to the northeast, and I needed to find my magical lens, my Wayne.

Old friends

I landed in Recife, another jumping-off point, a place on the coast with an airport, like João Pessoa where I'd been the year before and Maceió the year before that. The big cities were like beads strung along the highway that ran all the way south to Salvador, but for me the interesting roads were the ones that led in the opposite direction, away from the beaches, into the backlands. I hoped they would lead to some answers about water.

A song I heard on the flight from Toronto was still going around in my head: a jazz combo, the swish of brushes on cymbal, a jaunty piano, a crooner singing "Aren't you glad that you were born?" He had a luscious, teasing voice, and I pictured gleaming hair, winks, the clink of glasses and a conspiratorial roundhouse at the audience. The imagined scene stuck in my mind as the plane from São Paulo bumped down over the outskirts of Recife, the cardboard and plastic shanties and the turquoise ocean just glimpsed before we smacked onto the tarmac. The *favelas* were full of people from the interior, searching for a better life and washed up in a squalid fringe around the city. Recife looked big; it looked like all the rest. It was thirty-five degrees, and "Jingle Bells" was playing inside the airport.

My old friend Fátima was born in Recife. She had moved back from Maceió and there she was now at the airport, her long hair cut to a crinkly tangle. She wore a flowing red batik dress and two loops of heavy wooden beads around her neck. As we walked to her car she filled me in on developments in her life since we had last met: her move "back home," retirement, caring for an aged father who was married to a problematic second wife.

"*Mergulhar com cinto de cimento,*" she said darkly. Go diving with a cement belt — that was a new idiom for me. Being Fátima's step-mother looked like a challenge all around. She slammed the door, riffled through a magazine of CDs, slid one into the car stereo and stabbed at buttons.

"Buddhism," Fátima muttered. "Every day, I meditate. To keep me calm." She shot out of the parking lot, overtaking three cars and a motorcycle courier while I struggled with my seatbelt. The old car, the one with the leaking air conditioner and the one eye that I had driven in Maceió, seemed to have been replaced. It was good to meet an old friend in a new town, especially a friend with air con-ditioning. We stopped at a traffic light and a guy in a full clown suit complete with an exploding green wig shoved a leaflet through the window. He must have been dying out there.

"New car?" I said. Fátima nodded. Somehow I didn't feel safer. Her driving was the same as before: still the darting through rapidly closing cracks, the handbrake pulled on and forgotten when the light changed, the honks and imprecations, the sudden bursting into song. Had Fátima meditated today already? She didn't seem to have changed much.

On the ground, Recife looked even bigger. Most of the hotels are on the beaches, which extend for miles north and south of the city. They're like the beaches in big cities everywhere, or at least big Brazilian ones, featureless expanses of sand backed by a ribbon of cars and then buildings as far as you can see. On a whim, I opted instead for Olinda, the old capital of Pernambuco.

"It's quiet there?" I asked.

"It's quiet," Fátima said, gunning ahead of a truck that had two kids on bicycles hanging onto its rear end, catching a ride. "They say the best view of Recife is from Olinda." She waved at a hill in the distance. It took us half an hour to get there, and I felt guiltier by the minute for putting her to all this trouble, then slightly foolish when I realized the place had been preserved for tourists. The Polícia Militar wore blue armbands that said "Tourist Police;" they looked embarrassed too. I flagged down a group of four of them strolling along a cobblestone street.

"Where's the Hotel São Francisco?" Fátima had heard about it from a friend, and the name was certainly appropriate. "On the Rua do Sol," I added, making it even easier. They conferred for several minutes, then all four flung their arms out in different directions, like figure skaters. I found another place instead, on a quiet, narrow street that turned raucous that night, the shouts and the pounding of music pouring through the iron grillwork and filling the hotel courtyard. I should have known; for Fátima, "quiet" included loud music.

I went down there later to eat, at a tiny bodega across from the hotel. The bar was also a kind of general store, scaled down to the point where its offerings had become arbitrary, bits of this and that. Much of the stock was hanging from the ceiling: pink sandals, a thermos, an unvarnished violin. I took a can of beer and a cod-filled pastry out to the steps, watching the cars squirt by and the old couples emerge from their houses with their chairs. Next to me, a young woman drank beer from the bottle and sang into her cell phone; behind, inside the bar, three middle-aged men argued about their retirement benefits. An old woman, black and sticklike under a too-large baseball cap, meandered to a stop in front of me.

"You're beautiful," she said. I gave her one *real*. She smelled of urine. The crooner in my head sang, "Aren't you glad you're you?" When I went back in to pay, the three men had switched to telling *"Três mulheres"* jokes, the ones that start, "There were these three women. . ." I bought a custard and took it back to my room, eating it on the tiny balcony which was like a tiled pocket in the roof, looking past the massive dark shapes of the trees toward the lights of Recife. Fátima was right, it was a great view, even if something was biting my ankles. A breadfruit detached itself from one of the trees and whistled through the branches to thud invisibly into the ground, leaving a single leaf briefly suspended against the night sky.

The next morning there was a fresh breadfruit loaf on the breakfast table. I washed it down with strong black coffee and went with Fátima to see João Suassuna, one of the best-known and certainly the most productive of the critics of the São Francisco project. Suassuna works as a researcher for the Joaquim Nabucco Foundation, a big NGO that, oddly enough, gets most of its funding from the same government its researchers take to task so harshly. Suassuna is an agricultural engineer whose articles pop up everywhere, especially on the Internet. Since the massive increase in popular interest following Dom Cappio's first hunger strike Suassuna's face was becoming familiar from appearances on television. I'd approached him by e-mail months earlier and we'd progressed from Professor João and Professor Brian to some semblance of familiarity. I was looking forward to meeting him, armed with my usual wariness of "experts in offices."

Fátima and I found him at the foundation headquarters, at the end of a labyrinth of corridors inside crumbling buildings that together formed a large compound in the centre of the city. Tropical vegetation erupted everywhere, shading paths and walkways. I never did figure out what all the buildings were used for. One of them was very old, a colonial mansion with pink walls and white shutters. In Recife, Fátima told me, when newer buildings are put up on an old property the original house is often restored and dedicated to some related use — a social centre, say, in a new apartment complex. Maybe this one was reserved for entertaining more important visitors than I. Suassuna's office had smudged, tape-marked walls, a desk that was nearly obscured by reports and a computer, and a small conference table. One wall was covered in the ubiquitous Brazilian tiles, these ones white and making me feel as though I were in a bathroom.

Suassuna himself was tall, with the neat salt-and-pepper beard that so many middle-aged Brazilian men seem to cultivate. He looked oddly like Ciro Gomes, the Minister of National Integration and the strongest public promoter of the São Francisco diversion; they could have been brothers. Suassuna had a lot of energy. He talked rapidly, in the manner of someone who had said it all before. His right knee jiggled under the table. Even though his Portuguese was precise, with none of the smudging and dialect I found so difficult, taking notes was out of the question. Suassuna had too much to say.

"An *elefante branco*," he declared: a white elephant. That's what the diversion was going to be, and it would be a disaster like the Colorado, Lula's undoing. Suassuna cracked his knuckles and reached for a report. "And the quality of the water, that's another thing nobody talks about. But the São Francisco is already polluted, even the people living next to it can't drink it, so how can you say you'll solve problems by bringing it to the northeast? It's nonsense." He took me through the familiar numbers: flow rates, existing allocations, the miles of canals to be constructed over uncharted territory.

My pile of reports got higher but little of it was new to me and the pace of Suassuna's delivery started to become hypnotic. The effects of travel descended heavily, and I found myself thinking about the position Suassuna represented. Important, trustworthy and certainly indefatigable, he was all these things, but what happened to a person when they spent so much time in an office, analyzing, commenting, fulminating? The year before, when I had driven into the interior of Paraíba, the view from the highway alone had spoken more strongly than all the reports I had read, and I desperately wanted to head inland, away from the sea, to stop the car and get out and point to a dried-up pond or a brimming reservoir and ask my own stupid questions. Suassuna was a respected authority, and I deeply appreciated his taking the time to talk to me, but in the end he was no different from his reports, and I'd already read those.

I waited for the flow to falter and asked a final question. What about the water supply for Recife? Is the city in line for any of the water from the São Francisco, as Fortelaza, farther up the coast, will be? Not necessary, said Suassuna, because Recife sits above the band of sedimentary rock that lines much of the coast and, in contrast to the crystalline rock farther inland, contains enough groundwater for human use.

"But even that supply is being overdrawn," he added. "Too many wells, too much use by industry, too much seawater seeping in." I recalled seeing a bicycle that morning, on the way to Suassuna's office, trundling in slow motion at the edge of the fast-moving stream of cars. The bike was modified, fore and aft, to carry carboys of filtered water. Each one must have weighed twenty kilograms, barely liftable, and there were an astonishing six of them rigged around the rider, so that he looked like a raft surrounded by water.

Barbara Tuchman's criteria for folly include two that could have been written for the São Francisco diversion: that there be long-term and concerted opposition, and that there exist better alternatives. Her third criterion, that the proposed action be contrary to self-interest, was what João Suassuna was getting at when he said the project would be a white elephant, the undoing of Lula's government. For me, the most compelling of Tuchman's criteria is the existence of alternatives, because it's another way of saying, is this the right way to deal with the problem?

If you immerse yourself in a case like this one — see it, feel it, talk to people and read their millions of words — you come at last to deeply crave something more than polemics. All that negativity is enervating, and there's the nagging feeling that, in the end, the side with the biggest budget usually wins. Luckily, you'd have to be from another planet to miss the main point of nearly every critic of the São Francisco diversion project, namely that diversion isn't bad per se, it's just the wrong thing to do in the northeast. The right thing to do, critics like Suassuna say, is to explore and amplify the alternatives. The right thing is to live with drought.

I thanked Suassuna profusely and dropped a few more kilograms of reports into my bag. On the way out I stopped in the bathroom. When I pulled the string on the wall-mounted tank it gave a little cough and my urine stayed in the bowl. There wasn't any water to flush it away.

MY GLASS-BOTTOMED BOAT

The next day I got into my rented car and drove into the interior, determined to find something positive. The images had stayed with me from the last visit — all that water in reservoirs and *açudes*, all those tanks and pipes and pumps — and I had spent the intervening time immersed in polemics about water in general, and about the folly of the São Francisco diversion in particular. I'd had enough of argument. I wanted to see solutions. Heading inland from Recife wasn't that much different from penetrating Paraíba the year before, but it was still a shock to step out of the car into the alien landscape of the *caatinga*. The sun was palpable, like a weight descending on my head, and the dusty ground threw the heat back up so that I felt incinerated from both ends.

It was stupefying, this heat, and I thought, how else can you deal with a life in such an environment, if not through fatalism? And how difficult it must be to argue that the hardships were anything but the castigation of God. The sun drilled through my shirt and into my shoulders; it even *felt* vengeful. But I was here to learn about the water. I wasn't going to get very far just standing in the sun; the truth wouldn't be baked into me like a suntan. I needed help. I needed that expatriate, badly.

Her name was Beth Szilassy. Fatima told me about her in Recife: an expert in agricultural development, an English speaker who had lived in the northeast for twenty-five years and was, oddly enough, a Canadian. Beth could flatten my learning curve dramatically, shave months of research, steer me clear of blind alleys. She lived in Brejo da Madre de Deus, a town of thirty-seven thousand, about fifty kilometres down secondary roads into the fringe of the *caatinga*. Brejo sits beneath a big grey eyebrow of rock ("I've climbed that one, oh, maybe a hundred times," Beth told me later) and it's a pretty town, dating back to the construction of the first church in 1760. The economy has always been based on farming, mostly milk and carrots. Lately a new crop was being tried in the region, the castor bean, which has a future as a source of bio-diesel. Brejo is treed and a bit more humid than the surrounding *caatinga* (in Brazil, "brejo" means a small wetland), and at a thousand metres above sea level, the mountain behind it can get cool at night ("I take a sleeping bag," Beth told me).

Beth's directions were simple: three hours out of Recife, turn left at Caruaru, then left again at the big statue of Lampião. That part was pure "northeast;" Lampião was the bandit whose depredations in the early twentieth century put a price on his head, which was eventually removed for him. He was cut from the same cloth as Satan João, the bandit who found God and joined the doomed settlers at Canudos. Lampião was an unregenerate killer, and I can only explain his veneration as a way of recognizing the social standoffs in the *sertão* that made him what he was. Or maybe cultures just love to romanticize their bad guys. Al Capone, Lampião — what was the difference?

"It's a big statue," Beth assured me on the phone, "you can't miss it, right at the T-junction. But watch the turn, eh? You know about left turns here?"

Beth actually said "eh," the way Canadians are supposed to and nobody I know ever does, and she even did it when she spoke Portuguese. I told her I know about left turns and would try to remember: pull off onto the shoulder, look both ways and then dart across both lanes, wheels spinning. Lampião was there all right, the old bandit looming over a plain dotted with boulders, a giant's abandoned game of marbles. When I got to Brejo, I found Beth by asking a succession of people where the Farmer's Syndicate was; it turned out to be a new building, turquoise inside and out and built by foreign donors. It was next to an office providing rabies shots. There was a conspicuous Canadian flag painted on the wall of the lecture hall, so I guess I must have helped pay for the place.

The room was full of farmers in metal chairs, surrounded by posters and photographs. I had arrived in the middle of a training course, and the farmers might as well have been fishermen back in one of the towns along the São Francisco. Training courses are rural fixtures, on everything from engine maintenance to getting microcredit to novel ways of wringing income out of byproducts of farming and fishing. Beth popped up from one of the chairs, spotting me immediately. She tapped her watch.

"We should go, like *now*. There's a couple of cisterns being built near here, the guys were up at five this morning. Otherwise it's too hot, eh?" I had been up early too, to get here before the work was finished. Cisterns were what I wanted to see. I grabbed some sugary lemongrass tea from a table in the lecture room and we headed back to my car.

"Normally I use my motorcycle," Beth said. She lived in Brejo, but much of her work was back in the bigger city of Caruaru ("Capital of the Agreste"), forty kilometres away. I thought about her making that left turn, where the bandit's statue was, day in and day out.

Here was my magic lens. I couldn't believe my luck. Beth Szilassy looked to be in her forties, in brown jeans, T-shirt and sandals. Her brown hair was cut short and simple and she might have been a farmer herself. She'd worked for the Mennonite Central Committee since graduating from Guelph Agricultural College twenty-five years ago, all of that time in Brazil. Here in Brejo, her

Beth Szilassy and cistern clients, near Brejo da Madre de Deus.

biggest responsibility was cisterns. She talked fast and loud; being with her was like trying to take a drink from a fire hose. I stuck my head into the stream, telling her of my puzzlement about water storage in the northeast.

"I was showing some visitors from Hungary around out here, a month or so ago, can't remember exactly," Beth said. In Beth's life, a lot happened in a month. She gestured to bouldered plains, a few tiny mud houses with tiled roofs, a pond with retreating edges and a few cows nosing around the dried-out bowl. "They said, why are you building all these cisterns when you've got all this water around?"

"Exactly my question." I jammed on the brakes and pointed. "Like that one, that *açude* over there. Whose is it?" Water lay in a hollow, an expansive pond ringed by a shelving brown edge, a bowl left out in the sun. You could see where the stream had been dammed during the rainy season. Sir Richard Burton, when contemplating the effects of drought, envisioned the construction of "artificial lakes and reservoirs on the secondary streams." Well, here was one of them. There are more than seventy thousand *açudes* in the northeast now, most of them fairly small, like this one. Evaporation losses of up to 40 per cent mean that many are salty, especially the ones the owners proudly declare to have "never dried up." They never dry up because they're excavated in crystalline, impervious soil, about the same as building in concrete. Nothing ever trickles out; whatever solids are in the water simply get more and more concentrated. In a place like this, an *açude* was just like those crystallization experiments we did in elementary school.

"It belongs to the landowner," said Beth. "And it's polluted. So's the stream, so's all the water around here. Pesticides and fecal. See those cows?" A few cattle were poking around the edge in slow motion. "Lots of people still have to drink from places like that. It's pretty much guaranteed you're going to get sick. Especially the kids and the old ones." So: water, yes, but bad water.

"No sewage treatment?"

"Straight into the river. Back there, in the other direction from Caruaru? Remember a place called Toritama?" I recalled the sign, and wondering about the Japanese-sounding name. Beth went on: "The Capibaribe River runs through the town, goes all the way to Recife. Did you know Toritama is the jeans capital of Brazil?" I didn't. Toritama didn't look big enough to be the capital of anything. "They wash them, to get that faded look, eh? Over and over again, they've got these huge laundries just for jeans. It all goes into the river."

I had to think about that one for a moment. Pre-washed jeans, another of modern life's harmless inanities turning out not to be harmless after all. I imagined thousands of pairs of them, destined to be stretched across thousands of buttocks, tumbling in an immense and groaning washer the size of a swimming pool, and the Capiraribe River running away through the *caatinga,* faintly blue like an elderly lady's hair. Then what about the water in the tanker trucks, where did that come from?

"There's a place up in the hills, above Brejo. It's a bit cleaner. But then you depend on the municipality. You know about that?" I knew about that, how access to water was controlled. We had passed the *carro pipa* (tank truck) on our way out of town, parked in front of the municipal office. *Disk Agua,* it said on the oval back of the tank. Dial for Water.

"So there's water and there isn't water," I said. "Enter the cisterns, right?"

"That's the beauty of it," said Beth. "Turn here." We bucketed down another dirt road, the little Fiat rolling from side to side like a dinghy in a crosswind. Getting around on the motorcycle must be an adventure.

"Too late," said Beth. *"Damn.* But stop anyway, at least you can see the hole." We were outside a tiny house, whitewashed and sitting on dirt the colour of *café au lait.* The roof was tiled; the door and two flanking windows took up most of the front wall. A pile of gravel sat in front of the house and a crude fence made from two levels of barbed wire ran between gnarled sticks of brushwood. Above one window hung a wooden bird cage. A young woman, pregnant and wearing a striped top, leaned in the doorway. She looked to me as caged as the bird; neither reacted to our arrival. Above the red tiled roof, the northeast sky extended like the roof of a lavish diorama, brilliantly blue and pocked with photogenic white clouds, the kind that never mean rain. Where I live, this kind of sky lifted people's hearts; I had to remind myself that here, in the semi-arid, it was deadly. Things died under a sky like this one.

I went up to the door and introduced myself, looking over the woman's shoulder into the gloom. Curtains divided the house into rooms. A cheap boom-box sat on a table beneath a carving of Christ on the cross. Behind the woman I recognized a poster I'd seen in the Farmer's Hall, a checklist for maintaining your cistern. There were piles of white PVC tubing on the floor, and a fat roll of galvanized sheeting. "For the, um . . ." I couldn't remember the Portuguese for gutters; probably I never knew it.

"*Calhas,*" offered Beth, and the woman finally smiled. "They bend it into a trough." I remembered a *pousada* on the coast, in Ilheus; when it rained there, the water shot off the roof in parallel jets, like braids. Now, when the same thing happened here, that water was going to be caught and stored. The family's tile roof would be put to good use now.

"Can we see the hole?" I asked.

Concrete forms for a cistern in Pernambuco.

It was at one corner of the house, around the back, twenty feet across and maybe six deep. It looked like the kind of hole you throw bodies into. Here was the geology of the northeast, laid open as though for a textbook: the soil was gritty with small stones, so that the vertical edges of the hole looked like coarse sandpaper magnified thousands of times. A crude ladder poked up over the rim, bits of plank nailed between rough stringers. Digging would be slow and painful, even in the cool of early morning. Later, Beth told me about one of her clients, a young single mother, who had dug hers single-handed. It took her three weeks. The owner of this one couldn't possibly have. She was seven months pregnant, barely moving in the heat. Children drifted in, one on a mountain bike whose knobby tires were white with dust. They were all barefoot. We stood around in the pebbly soil and gazed down at some very low technology.

Inside the hole, a cistern was taking shape. Three concrete sections had already been poured, one on top of the other like African neck rings, bringing the cistern level with the surface. Each ring was made up of ten or so reusable wooden forms that together completed the circle. Lengths of metal rebar stuck up from the top ring like a crown. Another three or four circles laid on top of these and it would be done, half in the soil and half out. Beth told me a neighbourhood team could mount and pour one level a day, so that the average cistern could be constructed in a week. I got the feeling that, for this family, what was going on here was still a curiosity, something that was supposed to make their lives better, but for now was just an ugly hole into which the men crawled every morning and returned, two hours later, scratched and spattered with concrete and stretching before they left for the fields. We got back into the car and drove farther, to places where the cisterns had been operating for a year.

Here was different. In one house, a woman rested her elbows in a window framed by the network of white plastic pipe that collected the water from the roof tiles and channeled it into the cistern. She might have stepped straight out of rural Portugal or Italy, with fair skin and an easy smile. For her, the impact of the cistern was simple: no more disease in the family. The water was clear and with a splash of bleach it was bacteriologically safe. Not far away, an old lady emerged from her house to show off her own cistern; this one was sunk nearly level with the surface of the earth, with only the conical roof and a few feet of whitewashed wall protruding, like a yurt that had melted its way into permafrost. The official plaque on the side said this was cistern number 108,273. A pebbly plot of cactus next to it fed the cattle, the cuttings from already-harvested mature plants half-buried in the ground and looking for all the world like a miniature graveyard. The whole yard was fenced with brushwood sticks. A clay oven stood behind the house, next to a pile of brushwood for cooking. Across the road was an empty field covered only by a ragged brown furze. Its only feature was a pair of goalposts.

The old lady was wizened and nut-brown, in a white headcloth and a riotously coloured T-shirt. The metal hatch in the cistern was at waist height for her, and when I leaned over and peered into it I could see nothing at first, my eyes following the rope, adjusting from the glare outside. The end of the rope resolved itself into a bucket, then finally I could make out the surface of the water, surprisingly far away and lightly dusted, as though in a vase sitting in an old house. But it was clear. The bucket wobbled and the old lady drew it up and handed it to me. Its sides were cool.

"A gift from Jesus," she said, beaming. Then she let it fall again, and it made a hollow splash. I straightened up and she closed the hatch and we were back in the dry, bleached-out world of cactus tombstones and dun-coloured dirt.

Gift of Jesus, gift of the government — what did it matter when the cistern offered such a dramatic alternative? For this *nordestina,* the climate was God's castigation and the cistern was His blessing. What mattered most was that she didn't need the *carro pipa* anymore. She didn't have to lug eight kilograms of water every day — which, by the way, is about the same as a couple of four-litre jugs of milk, except that it's not so conveniently packaged — and her family could afford to use more than the equivalent of one toilet-flush worth of water a day. What mattered was that, down there beneath her feet, there was cool, clean water.

It was like going back in time, to when you really could drink the water of the São Francisco, as Richard Burton had. In a trip of four months, the man was never sick, something he attributed to drinking the water of the river (all you had to do was settle it first), coffee to keep up the "vital heat," lime juice against scurvy, cognac or *cachaça* at morning and night, and two grains of quinine to ward off malaria. Walk and talk were parts of the hygiene; above all "activity of mind, plenty to do, and contentment." Well, maybe this woman and her family were getting some of that back.

We headed back toward Beth's house in Brejo, and already the landscape had subtly changed for me. The *açudes* made sense now, and so did the little piles of gravel outside houses. I asked Beth how many cisterns had been built so far. "Just over . . . let me think." She snapped her fingers in that Brazilian gesture I could never master, urging the words out. ". . . Just over a hundred thousand. I've got it all on my computer at home, the funders, the way we choose the families, the training courses."

I worked it out quickly: at a thousand reais per cistern the million cisterns would be a round billion. Probably a fifth, more likely a tenth of what the transposition would end up costing, with its hundreds of kilometres of canals that would pass by houses like these without stopping, water from the São Francisco vanishing literally into the air. Cisterns were clearly a deal. They were cheap, ridiculously so by the standards of foreign donors looking for concrete results. You could build three of them for the price of a plasma TV. And, like so many overlooked solutions to enormous problems they were simple, like the mosquito nets that can do as much to help eradicate malaria in Africa as ten times their cost in vaccine research, or the little packages of salts that help in rehydration after diarrhea. Their footprint on the environment was invisible.

But what was beginning to dawn on me too was that cisterns were also subversive. The simple act of sinking a hole in the ground and diverting rainwater into it allowed a family to cut cleanly through the umbilical cord of obligation and favours that symbolized everything that was socially wrong about the northeast. A gift from Jesus, maybe, but an undeniable emancipation as well. I thought about all the social programs I had seen in the developing world, the empowerment workshops, the gender sensitizing. Maybe these holes in the ground accomplished more than all of them. Another cistern went by, well back from the road but catching the sun like a precious stone.

"This must be the cistern capital of the northeast," I said.

"Oh no. There's lots more of them the farther inland you go — remember, I just handle one micro-region. But hey, Brejo is still the capital of *something*. Turn here." We were back in the town, and Beth was pointing to a statue I'd missed before, in the middle of an intersection. *Monumento ao Toyoteiro* it said, above a life-sized sculpture of a Toyota Bandeirante bursting through a wall. As though to emphasize the point, there was a real Toyota parked next to the monument, its hood propped open.

"You mean, all those Toyota trucks I've been seeing, those weird big ones . . ." Now that I thought about it, there had been a lot of them on the road from Caruaru, mostly full of passengers, the roofs piled high with boxes and sacks. One of them had almost hit me head-on, yawing into my lane for a heart-stopping second, a door to the infinite suddenly opening: *Now? Not now. But maybe later.* A

white one, with a gleaming chrome grille. Named for the marauding Portuguese settlers of four hundred years ago.

"Right here in Brejo," said Beth. "They strip them down and
stretch them, turn them into minibuses, eh? Brejo is full of little
shops doing it. There's more than three thousand of them. in
Pernambuco. Want to see one?"

In three minutes I was shielding my eyes inside a gloomy warehouse while two guys crouched in a shower of sparks inside a gutted
Bandeirante, welding its new mid-section into place. The new passenger area was hooped with three sturdy roll bars, and there were
a dozen other bodies nearby, in varying stages of transformation. In
one, a gleaming new Mercedes Benz engine had just been inserted;
another three had just been painted a uniform primer-grey, various
smaller parts dangling on strings from one of the roll bars, like
Christmas ornaments. The upholstery division was a cluttered
corner with two ancient sewing machines and a littered workbench;
the floor was strewn with fabric cuttings and a set of beaten-look-
ing cardboard patterns hung from nails driven into the brick wall.

I thought about the Toyota that had almost hit me on the road
to Brejo and wondered what kind of safety checks went on in a
place like this. In development-speak, places like this would be
called micro-enterprises, but I doubt many development bureaucrats would take a ride in the product. But that was the whole point
about development, wasn't it — to add a little to the value of human
life? Here in Brejo, where $500 brought you clean water, life was
still relatively cheap.

I picked through more cistern files in Beth's house, popping her
organically farmed strawberries into my mouth while she fired up
her laptop. It sat surrounded by piles of reports on a scuffed desk
beneath a large painting of St. Francis, patron saint of the São
Francisco. The walls were the deep blue of the sky a *nordestino* sees,
day in and day out. A narrow single bed occupied one corner;
bookshelves were crammed with agricultural reports. Beth had
records of everything related to the cistern program: maps, family
demographics, costs and donors by year, meetings. She showed me
the official documentation for each family, right down to the GPS
coordinates and a digital photo of the family with their new cistern.

One set of forms was the questionnaire filled out by the prospective family; the list of diseases "most frequently experienced" ran to fifteen, and most were water-borne.

FATE AND PEANUTS

I drove back to Recife thinking about holes in the ground. An image began to form in my mind, like the view from an airplane as it comes in to land in Palm Springs, say, or Los Angeles. There, a backyard swimming pool stands out like a diamond in a navel, and each one is attached to a collapsing aquifer beneath, or a gasping river thousands of miles away. The image of the northeast was different; instead of blue, the studs were a dazzling white, and the water they contained wasn't stolen or bartered but it had the power to give life and to break a centuries-old chain of dependency. It was a very nice image, and all for five hundred dollars and a week's work.

I felt I'd found something important, that my search for the positive in the haystack of negativity about the diversion hadn't been in vain. Who knew where it might lead? In Japan, I remembered reading, citizens of Kobe were snapping up government-subsidized fibreglass cisterns to store rainwater in the event of another catastrophic earthquake; a single two-hundred-litre tank cost almost exactly the same as these sixteen-thousand-litre Brazilian monsters. My imaginary friend Miss Tojo was the worrying type; she probably had one.

Twenty kilometres out of Brejo, at the intersection Beth had warned me about, a crowd had gathered. As I inched past, more cars pulled over, doors opened, people hurried over at that peculiar half-run that reveals both excitement and dread. A few Polícia Militar were standing on the fringes, and as I got closer I could see into the centre of the circle, where one of them seemed to be guarding a sheeted figure. A twisted motorcycle lay a metre away, and a thick double trail of blood ran out from beneath the sheet and disappeared under people's feet. The blood was dark and shiny and recent. If I had left Beth's house five minutes earlier, I might have seen the rider gun too late across the intersection, wobble and smack.

Or not; perhaps the alteration of my movements might also have affected the machinations of fate, and the dead man might be in Caruaru already, climbing off his machine and stretching. It depended on whether you believed in fate, in the hand of God, in punishment. Probably most of the onlookers did — why else would they congregate here if not to verify for themselves that they were not dead, that this poor guy was, that life went on? God gave one old lady a cistern, and He took this man's life away. I rolled past, watched by the bandit Lampião, then pulled back into traffic. Lampião himself wouldn't even have blinked, and it was doubtful whether God meant much more to him than the certainty that he had long been damned, and that he might as well just go on adding to the debit side of his ledger. I thought about Beth on her motorcycle, how she turned left at that intersection every day. The little tableau dwindled in my rearview mirror and disappeared.

But the accident stuck in my mind: the blood, the crowd jockeying for a better look, the passive police. Three ambulances flew past, coming from Caruaru, a strange response for a dead man. It was hard not to feel fatalistic myself. Of course the diversion would go ahead, I thought, in a year or in ten, just as there would be more motorcyclists killed on this stretch of road. There wasn't any mystery about why the accidents kept happening, any more than about what was wrong with the diversion. Maybe the sudden, violent death on the highway was not so very different from the hunger strike of Dom Luíz Cappio: it attracted a crowd, people talked excitedly for a while, then got back into their cars. Someone in authority would have to comment — agree to look into the reason why so many Brazilian drivers get killed while turning left across three lanes of traffic, promise to hold further discussions on a $5 billion project — but life would go on. I remembered something Beth had told me, describing the murky system of land ownership.

"I went into the land reform office in Caruaru last week," she said. "The place was empty, just a secretary, that was it. I couldn't get hold of any of the title documents because anybody who knew anything was up in Cabrobó, doing surveys. Same thing in all the other land offices in Pernambuco, everybody's in Cabrobó." Cabrobó was where the hunger strike had been; it was also the place where the first canals would be built to suck the first water out of the São

Tempting fate near Patos, Paraíba.

Francisco. Surveying meant expropriation, construction, progress, all of it on the very site of the biggest popular manifestation against the project. It wasn't exactly what the land reform offices had been set up to do, which was to help along the painful process of transferring ownership of land to families who had been squatters for centuries. Life was going on.

I turned on the car radio. Somebody was singing about the São Francisco, how it was dying, a litany about disappearing fish, pollution, deforestation. And then, "God made it so, so nobody can complain." More fatalism, more bodies by the side of the road.

Back at the hotel, I took up my usual place on the curb and ate a *pastel* from a nearby vendor, washing it down with beer from the minibar. The pastry was just right, not greasy like the versions in the shopping centre food courts but hot and crisp, without a trace of oil from the deep fryer, the chicken and vegetables inside tender and spiced. I bought another one and returned to my spot, staring at the door of the black SUV parked in front of me, and at people's knees

349

as they squeezed past. Another vendor sold some peanuts to the man sitting next to me, three twisted paper cornets warmed by a little portable brazier that he tended there on the curb, picking out dead coals, stirring the living ones.

I accepted some peanuts from my neighbour and took them back inside to the hotel computer, where I logged on to the Internet. There was more news about the river: the Brazilian Army Corps of Engineers was getting into the act, with teams of military surveyors working the territory around Cabrobó, and also around Petrolandia, where the other wing of canals would start. These must have been the guys Beth told me about.

"Not to worry," the spokesman said, "just some government infrastructure work, nothing related to the diversion." It was a typical harvest of diversion news, all of it depressing. Meanwhile, off to the side and unnoticed, villagers throughout the northeast were doing little water projects of their own, taking delivery of bags of cement, piles of gravel and rolls of galvanized tin, getting up at five in the morning to dig holes in the ground so they could divert a little rainwater. I clicked on and ate the warm peanuts, drank some more beer. Fate, it seemed, was moving along as planned, at least as far as the diversion went.

But in other Brazilian news, life went on. The ex-heavyweight champion Mike Tyson was in São Paulo, his first visit to Brazil. Now here was something to take my mind off water. Things had started shakily; early in the evening Mike had handed a hundred-dollar bill to a beggar, probably ensuring the kid a sound beating later, if not death. Later, a turn for the worse: on returning from a nightclub with six women in tow, the champ was told they would not be allowed to accompany him to his chambers. He was forced to eat alone, consuming, the article related, *feijoada,* rice, steak and hot sauce. I imagined him in his room, still dressed in the black leather jacket and white running shoes he wore in the picture on the monitor in front of me, testosterone fizzing, teeth grinding. He was hunched over the mound of glutinous beans, shredding the steak and muttering, his mind clouded with images involving six women and a desk clerk. His jacket alone probably cost more than a couple of cisterns.

I climbed the stairs to my room and turned on the TV: more ridiculous images. There are normally four broad classes of TV programming in Brazil: *telenovelas* (soap operas), news, variety shows and football. None of them is especially relaxing, although I have found that turning the sound off on a soccer game produces a colourful, swirling effect, like a screensaver or one of those early computer games where tiny figures ran pointlessly around. Tonight it was a variety show featuring a grinning panel whose topic was the application of rejuvenating skin treatments.

Three members of the panel sat on a sofa while the fourth circulated between them and a girl in a bikini who was being laboriously painted by a stern middle-aged woman in a white smock. She reminded me of my first piano teacher, Miss Calvert. The girl was pretty and very young, with traces of baby fat around the straps of her *tanga*. The host was thickset and mustachioed, in a dark double-breasted suit that emphasized his bulk. He wore a headset and oozed the kind of masculine confidence that has long disappeared from North American programming. He loomed, peered, pointed with a stubby finger, and the camera travelled lovingly over goose-pimpled flesh turned green and glistening. Miss Calvert dipped and stroked, and the girl smiled bravely. It seemed to be an awfully small brush, so it was taking a long time to cover the girl.

It had been a long day. The drive to Brejo and back, the cisterns, death and fate, beer and peanuts. I thought that, if I were given the job of painting the girl I would use, at the very least, a good quality three-inch bristle brush, possibly even a spray gun. I imagined that somehow Mike Tyson had joined the panel. The others made room for him nervously, scooting to the corners of the sofa. He was scowling and still wearing the leather jacket and white running shoes. There was dried *feijoada* on the front of his white T-shirt. The portly host hurried over and stuck out his hand.

"Welcome, Mr. Mike! Welcome to Brazil!"

"What is this Mister Mike shit? I ain't no cartoon character."

"Yeah, okay, okay! We call you *Campião*, what's that in English —"

"The Champ!" says a panel member. It comes out "chimp."

"Okay, yeah! The Chimp!" The host makes a conductor's motion with his arms and the audience applauds.

THE END OF THE RIVER

"The *fuck?*" Mike Tyson rises, brushing beans off his front. He pulls back a huge fist to clobber the host and catches sight of the girl in the bikini. My old piano teacher has finished painting and is now wrapping strips of white cheesecloth around her midsection. "*Oh* yeah." He lowers the fist and starts toward her.

"Now Mr. Mike, heh–heh, Chimp . . ."

The Champ places an open palm in Miss Calvert's face and shoves. She flies backward, overturning a bucket of green paint. The girl tries to make a run for it but Mike catches a corner of cheesecloth and she begins to unwind, spinning rapidly. The host tears off his headset and flings it at the Champ as a huge hand closes in on the camera and the screen goes black.

Maybe it was time to go to bed. I turned off the TV and reached for my copy of *The March of Folly*. Mike Tyson versus Barbara Tuchman? No contest. The Champ was on his back already.

THE GUMDROP EFFECT

The idiocies Tuchman related, blow by blow, were all political ones. Yes, they caused incalculable human suffering and the collapse and rebirth of states and even religions, but these effects were relatively short-term. Things blew over. Nobody gives much thought today to Montezuma's tragic lack of judgement. In her discussion of the folly of the Renaissance popes, Tuchman points out that the historical significance of the bloody "Italian wars" that led to the sack of Rome in 1527 is "virtually nil except as a study for the human capacity for conflict." The mistakes of Vietnam, which is the last of her case studies, surface only infrequently in Hollywood movies, and world leaders seem bent on repeating them.

But folly concerning the natural world, while stemming from the same human defects of pig-headedness and greed, is surely much more long-lived. Is the folly of the northern cod, of the Japanese salmon hatcheries, of the São Francisco diversion somehow hardwired into human behaviour too? Fisheries is just one arena in which humankind causes irreparable harm, eliminating or transferring species, simplifying ecosystems, tearing up habitat to inflict damage our children's children will be reminded of every day. The most colossal natural folly of all, climate change, will make the self-absorption of the Renaissance popes, the bloody scramble for Italy's flailing body and the opening of the door for Martin Luther look like tea-time gaffes.

These were dark thoughts to be having alone, in a hotel room far from home. The night was still deliciously warm, and there was plenty of happy noise from the street. I could have rung up Fátima, gone for a stroll along the cobbled streets of Olinda, organized a late dinner to sit on top of the *pastels* and beer; such things were not unheard of in Brazil. But I stayed where I was, trying to think it through.

It was really bothering me, this idea that depletions of natural resources are a special case of folly, fundamentally different from ill-fated conquests and plain bad politics because they involved the basic inability of humans to stop doing something that feels good. Call it the gumdrop effect. A bowl of gumdrops sits on the table; you eat a couple, promise yourself to stop there. But you keep passing the bowl, day after day, and finally they're gone. You know they're not really good for you, but they're there, so you eat them until they're not there anymore. Having someone nag you about your habit doesn't have much of an effect, other than to make you feel guilty. But guilt isn't usually enough of an inducement to change behaviour.

Here is Herman Melville, in *Moby-Dick*, on what was already happening to the great whales at the beginning of the nineteenth century: ". . . the moot point is whether Leviathan can long endure so wide a chase, so remorseless a havoc; whether he must not at least be exterminated from the waters, and the last whale, like the last man, smoke his last pipe, and then himself evaporate in the final puff."

Melville was mostly right; nations *did* slaughter away to the last drop of spermaceti — to the last gumdrop. So too, perhaps, with the *surubim* in the São Francisco and the salmon in the Fraser: the evidence that the bowl will one day be empty doesn't seem to register until it's too late, and communities have gone through the incalculable hardship of walking away, finally, from their livelihoods. To pace the docks of what used to be a thriving "Fisherman's Wharf" in British Columbia is to pass over a grave, except that some of the corpses haven't even had a decent burial. Back in town there might be a new hotel with signs advertising whale-watching for the tourists, but down on the docks is where the casualties are. A derelict troller, its never-to-be-renewed licence fading in the wheelhouse window, mildew growing over the radar dome, the deck littered with tangled line and the hull trailing a shroud of brown weed. Every five minutes the automatic bilge pump coughs out a few gallons of rusty water; the boat's on life support. There's no For Sale sign, because nobody's buying.

But the whales came back, slowly enough that an entire industry had to find something else to do for a hundred years while the numbers rebuilt. Maybe the northern cod will come back too, now that it has become uneconomical to chase them anymore; the same with salmon and probably a host of other examples, although the havoc that climate change is starting to wreak with aquatic ecosystems makes this kind of prediction a tough call now, one folly compounding the effects of another. I thought back to a scientific conference I had attended just before coming to Brazil, on the effects of climate change on salmon.

One by one the speakers took their place behind the podium, the big screen behind them filling with charts of rainfall and snowpack and river temperatures measured over decades. One presenter was bald and and stooped, in a badly fitting sport coat. The light from the lectern reflected off his glasses. He looked innocuous, but the message he was delivering was grim.

"We're seeing now what was predicted a decade ago," he said. Snowmelt-dominated river systems were now controlled by rainfall; air and water temperatures were rising so that, by the end of the century, river temperatures would *always* be in the danger zone for spawning salmon. Higher temperatures meant low summer river flow, which meant even higher temperatures. It was a vicious circle.

"If people don't trust the scientists," he said, "here's some *real* data. There's a betting pool in Dawson City, been going for decades. They bet on the date of spring ice breakup. And guess what? It's coming ten days earlier every year." The big question for science now wasn't whether higher temperatures were a fact; it was what to do about it, and short of absurd rescue attempts like redirecting spawning salmon into colder holding areas, survival was pretty much in the hands of the fish. It was all in the genes: different stocks were adapted to different river temperatures and now that temperatures were on the rise, the big question was, can they adapt? One speaker said, "We're watching a gigantic experiment in evolutionary selection." The conference closed with a perky address from the Minister of Environment, who congratulated himself for buying a Toyota Prius and announced a niggardly $230,000 in research grants for studying climate change.

"We must work to adapt," the Minister told us as we picked at our tiramisu.

How late is too late?

Did humans only ever learn by absolutely emptying the bowl? Did everything have to go the way of, say, the Murray River cod of southern Australia, which was fished into unprofitability by the 1930s, recovered slowly until it could be fished again, then succumbed to the combined effects of renewed fishing, dams and introduction of competing species? This was an intensely cynical thought, and it didn't make me feel any better about having spent so many years of my life trying to snatch the bowl of gumdrops away, or hide it, or wag my finger about the dangers of eating too many.

I got another beer, fumbling my way to the minibar in darkness. I didn't really want to drink it, but there was something satisfying about resting your elbows on warm roof tiles while holding an ice-cold can. Anyway, this was Brazil; beer was cheaper than bottled water. Someone had calculated that to supply every family in the northeast with bottled water would still be cheaper than diverting the São Francisco — but of course, supplying families with water wasn't really what the diversion was all about. So I was back to water again. If cynicism wasn't an appropriate response, what did one do about such folly?

I decided there were two valid approaches, and that the second of them worked even if you were a cynic. The first approach was to fight, to make jihad on the project. In Brazil you were surrounded by crusaders against the diversion (and against a lot of other things as well). João Suassuna in his bare office in Recife, with his statistics and his mile-a-minute delivery and his knee that never stopped bouncing, was a good example. So was Dom Luíz Cappio, perhaps the best and most powerful crusader, because he had not only the weight of undeniable experience and commitment behind him, he had God on his team as well. These two men were throwing everything they had at the enemy. They were eating, sleeping, breathing the folly of diversion, the ones trying desperately to get government to make the choice, as J.K. Galbraith put it, "between the disastrous and the unpalatable."

On the other hand, there was Beth Szilassy. She represented the other approach, the one that works no matter how cynical you are about humans and their governments, because it has the inestimable advantage of a real product, something you can step back from and dust your hands off and say, "I did that." Okay, Beth seemed to be saying, the diversion is absurd, people are being asked to believe that poor *nordestinos* are going to get the water they need and, by and large, they're not. But what if I can do something completely different, something that actually accepts the possibility that that particular bowl of gumdrops *will* be cleaned out? What if I help build cisterns?

By her actions, Beth was saying to me, "Let the technocrats and the zealots and the saints fight this one out to its — probably foregone — conclusion; I'm going to pour some concrete of my own." It was a philosophy that made a lot of sense to me, especially at the end of a day that included not only the constant reminders of the supreme folly of the São Francisco diversion, but also the graphic demonstration of how a simple technology like cisterns could profoundly affect lives. Even without the images of violent death on the highway and an ex-heavyweight champion running amok.

I fell asleep that night reading Barbara Tuchman's final chapter, the one that resonates most for me, maybe for the simple reason that I lived through the period when it happened: Vietnam. I opened the book and there they were, all those familiar, freighted names, the ones I saw in my morning paper through much of the sixties and seventies. Nixon, Westmoreland, Johnson, General Thieu. Even General LeMay, the architect of the firebombing of Tokyo in 1945 that killed a hundred thousand civilians, back for more in Vietnam. As usual, Mrs. Tuchman started off with a bang.

American policy, she wrote, had been folly from the very start, in 1945, when they decided to help the French enforce De Gaulle's hard line on Vietnamese sovereignty. His top general was already saying it couldn't be done, and it couldn't. Not then, not twenty-five years later. The French saw "their" stronghold in Southeast Asia divided in two, and America lost the second half for them twenty years later.

"Folly lay in attaching policy to a cause that prevailing information indicated was hopeless ... support for Humpty Dumpty was chosen instead, and once a policy has been adopted and implemented, all subsequent activity becomes an effort to justify it." That "subsequent activity" *was* the news for two decades, as more Americans and Vietnamese were killed, and more took their places. America opted for mass self-hypnosis and became lodged in the trap of her own propaganda. To me, the "domino theory" of communist expansion and the Brazilian government's "water for those who have none" were both examples of hypnosis, of rhetoric becoming doctrine that was systematically shoved down the throats of local "beneficiaries" who knew it was hogwash. Seen this way, my fear that the São Francisco "problem" was insoluble made even more sense.

This stuff was almost painful to read. Not just because it brought back memories, although it certainly did that. Johnson ignoring the advice of his Departments of State, Defense, Joint Chiefs, even the CIA, instead giving the order to pour the bombs onto "that raggedy-ass little fourth-rate country" — that memory came back as though it had happened yesterday. The really painful part was Tuchman's relentless dissection of the mindset, the group-think that allowed people to go down such a disastrous path. Everything that was happening in the northeast fit this pattern, every new e-mail bulletin from João Suassuna another piece of sensible counsel ignored by leaders marching firmly in the direction of folly. Tuchman's writing was brilliant; on nearly every page, a sentence would pop out: "The Administration simply went forward because it did not know what else to do." Or this one, on the futility of fighting a guerrilla war on foreign territory: "a tank cannot disperse wasps." Johnson's inner circle were "like men in a dream."

How clear it all seemed, from Tuchman's perspective of fifteen years (her book was published in 1984); another twenty made the blunders and sleepwalking look even more avoidable. Of course that was her point, that debacles like Vietnam and the Trojan Horse were avoidable, that the pathway to folly could actually be avoided. So could the foolishness of the São Francisco diversion, or the depletion of fisheries.

There's usually a moment when you can pull back. When the American bombing campaign "Rolling Thunder" began in 1964, the opportunity for what UN Secretary-General U Thant called "graceful exit" would never come again. Rolling Thunder was supervised directly from the White House, so detachment from reality was nearly complete, but it was almost a decade before the war finally ended. Was the deal that ended the first hunger strike on the São Francisco an opportunity for the diversion project to pull back too? If so, it was an opportunity lost. Of the rightness of the domino theory and the certainty that United States "weakness" in losing Vietnam would automatically lead to a Third World War, Johnson was "as nearly as anyone can be sure of anything." Well, it didn't happen then, and it still hasn't. The proponents of the São Francisco diversion are just as sure that all those new canals and pumps and reservoirs will bring prosperity to the northeast.

I put the book down, turned off the light and lay listening to the thump of music and the skirls of laughter from the bar across the street. The next morning, a few hundred kilometres inland, men would rise early, rub the sleep from their eyes and go down into cool holes in the earth, building their own simple emancipation from the unfairnesses of climate and custom. This is what Barbara Tuchman has to say at the end of her chapter on Vietnam: "No one is so sure of his premises as the man who knows too little." Maybe the men and women in the holes in the Brazilian northeast, with their shovels and bags of concrete, know a little more.

CHAPTER 17

"I GET THE SCIENCE"

Asking science what we should be asking ourselves

The real endangered species: a local fisherman and his sons near Três Marias, São Francisco River.

A HUNDRED KILOMETRES FROM WHERE I LIVE on Vancouver Island, there's an elderly man, Merv Wilkinson, who has been cutting trees in his own woodlot for sixty years. He's world-famous for proving you can take some, leave others, still make a profit. But not a fortune, and that's the reason the big logging companies won't — can't — follow his lead. Just as close to my home, there are any number of small-time salmon fishermen capable of making a decent living by fishing selectively, in the approaches to rivers, rather than lining up farther out to sea to form part of the gauntlet of nets that currently greets migrating salmon in British Columbia. A single gillnetter is the equivalent of Wilkinson in his woodlot; when you see one alone on the ocean, waiting at the end of the invisible curtain strung out behind it, all you see is one tiny boat and, a few hundred metres astern, the orange ball that marks the end of the net. It's like a hand thrust into a mountain of beans — how many can you possibly pick up?

But lone boats are like lone loggers: you don't see them anymore. I counted gillnetters once, during an opening in Bentinck Arm, near Bella Coola, and there were more than sixty of them, end to end across the fjord up which the migrating salmon had to swim. It was a gorgeous gauntlet, the cliffs walling the fjord like green buttresses. The fishing continued until after sunset, the last few boats hauling in and turning for home as the hills on either side darkened and disappeared. The fleet unloaded through the night, each boat waiting its turn to come alongside the buyer's wharf in Bella Coola, the crew standing around in oilskins and smoking while a suction hose rooted in the hold, feeding greedily.

There's a name salmon biologists use to describe the impact of a fleet this large; they call it the "boxcar effect." Putting that much fishing power in one place at one time actually creates a hole in the salmon run, one you can measure in space and time. It's like an enormous hand reaching down and plucking a container full of new Hyundais off a train. Attitudes about "resources" and "extraction" — two words that send chills down a conservationist's spine — don't seem to be place-specific, and the Brazilians who want to irrigate the northeast aren't much different from Canadians who keep sending out armadas of gillnetters to head off the salmon, or continue clear-cutting forests in British Columbia when sustainable, but much more modest, earnings can be had from eco-friendly practices like selective logging in small woodlots.

What's really at work in the Brazilian northeast, shorn of the underlying complications of fatalism and historical land ownership, is a clash of conceptions about what the land can, and should be, used for. It's a generic difference in mindsets, fundamentally no different from any other resource conflict. If you're a politician or legislator, land is an abstraction, something that exists in charts and diagrams and proposals. Land is not your world; your world is offices and meetings. Land is glimpsed from the window of an airplane, at which remove it seems reassuringly limitless. It's almost impossible to think small from thirty thousand feet, and the idea of growing drought-adapted crops for local consumption, the very idea of "coexisting with drought," has no intrinsic value for a politician. It delivers few votes. If you have a proposal in your briefcase to quadruple agricultural production and tilt the balance of payments in the right direction, all by diverting a little "excess" water, who in his right mind wouldn't support it?

Why aren't there more selective fisherman, more *nordestinos* saving rainwater and harvesting the fruit of the *jaú* for the simple and irrefutable reason that they know the tree has always survived in that inhospitable land, and always will? Why does greed always gain the upper hand? Maybe it's just human nature, what Carl Sagan calls the "shadows of forgotten ancestors" falling across our paths. It's fashionable to say that indigenous societies always had an innate respect for the land that allowed them to live lightly on it, taking only as much as was needed for subsistence, but there may be another explanation: they simply never had the technology to take more than the tiniest fraction — that handful out of the mountain of beans — of what nature had to offer. In the rare cases when they did, there is evidence that indigenous societies had no qualms about emptying out a particular location, because there were always new locations where the bounty was untouched.

Conquest and industrialization have had the same effect everywhere, bringing methods of extracting resources that made fortunes almost overnight. Making fortunes is habit-forming. There are more fortunes to be made from diverting the waters of the São Francisco, just as there were two centuries ago for planting monocultures of sugar cane and coffee. The addiction to extracting every last tree, fish or drop of water is centuries old, and I don't see it being overcome other than by the unpleasant ritual of going cold turkey. Even then, it will have to be an enforced withdrawal, not a voluntary one.

Conservationists are the world's folly specialists. What conservationists really do is identify one specific kind of folly, the persistence in gnawing away at a natural resource until it's crippled or extirpated, and they then devote their lives to trying to tell the perpetrators and the world that it's a bad thing. If Barbara Tuchman had lived into the 1990s, what would she have made of a notorious natural collapse like the northern cod fishery, which was completely foreseeable, even predicted, yet allowed to happen? Surely that was folly too?

Shortly before one of my trips to Brazil, I had lunch with a man who had been Canada's minister of fisheries during one early stage of the northern cod's slow decline. Many ex-politicians have the whiff of lost power about them, latching on and lighting up the vote-getting smile that's become an unshakeable habit. John Fraser is different; there's no smell of the has-been about him. He's in his seventies now, working harder than ever to save salmon, still courtly with waiters, as though being in their restaurant were a privilege.

"I'd like a Diet Pepsi, please," he said to the man serving us, holding up the wine glass. "In this one, if you would." John has always said this, in every restaurant we've eaten in. And then, as sometimes happens, the conversation turned to the past, and we talked about the northern cod.

"The scientists came to me with their data," he said, his voice developing an edge to it. "It was pretty clear, the average size of the fish was getting smaller. I mean, you didn't need a Ph.D. to figure out what *that* meant. Ah, thank you." He took his goblet of Pepsi from the waiter, interrupting his story to smile and make eye contact.

"The fish were getting smaller," I prompted.

He glared, as though the damning columns of numbers were brand-new, not twenty years old. "It was so damned *obvious!*" He balled his fists and swung them on either side of his shrimp salad, not hitting the table, but close enough. "I couldn't get anyone in the department to listen, especially not the managers who could make decisions. So we just kept fishing." He slumped back in his chair. "A *lot* of people knew something was wrong."

In my imagination, Barbara Tuchman had joined us. She was having the lobster. She put down a claw, wiped her hands on her napkin and leaned toward me. "The managers were displaying cognitive dissonance," she whispered. "That's psychologist-speak for 'Don't confuse me with the facts.' It assists in wishful thinking."

We knew the rest of the story: two more ministers thrown to the wolves, the fishery grinding to a squalid halt. "Cognitive rigidity," nodded Barbara, clanking a lobster claw against the plate. Ten years later, the moratorium wasn't worth lifting because the cod stocks that had once been so plentiful still showed no sign of even the slightest recovery (they still don't). Government had acted contrary to its own self-interest. Folly had prevailed. It was the worst fisheries collapse in recorded history. Mrs. Tuchman had the last word. "Common sense," she said, "is a faculty that has a hard time surviving in high office." *The March of Folly* is, not surprisingly, one of John Fraser's favourite books.

DOING THE IMPOSSIBLE

The science that Minister Fraser desperately wanted his managers to pay attention to may not have been perfect. By the time of the cod's demise, governments controlled research; science was a long way from the innocence of T.H. Huxley's era. The honourable discipline I was trained in had become cheapened, and the loss of virtue continues.

"Hey," the Canadian minister of the environment said about climate change in early 2007. "I'm not a member of the flat-earth society, you know. I *get* the science."

No, I thought, you don't. You're just watching the tide turn, and you know which boat you want to be seen in. If you don't like "the science" six months from now, you'll just tell your scientists to study something else. And they will. When the issue is climate change, all bets are off anyway. Climate change is the cocked fist that will flatten — make irrelevant — our niggling little worries about conserving this fish or diverting that river, because the very environment of the fish and the river is no longer a given. It's going to change, and keep on changing. That makes the science the Minister so confidently "gets" a special kind of science, one that relies less on hypothesis and experimentation, and more on prediction of future events based on complicated mathematical models. To be fair, I don't envy *any* minister's having to get *that* stuff.

I know a fisheries biologist, a short, affable guy with a lopsided mouth that makes him seem to be smiling at some private joke. In my friend's case (I'll call him Ken), the joke might be this: his job is to be a scientist, but his masters in the Fisheries department only want him to do half of it. Instead of answering a question the right way, by creating a hypothesis and testing it through experimentation, Ken only gets to do the first part. It works like this:

Management: "Ken, we need to know whether building a humongous coal mine on Immaculate River is going to adversely affect the population of Barrett's lesser sculpin. You know, that ugly little thing that's listed as endangered."

Ken: "But there aren't any data on Barrett's sculpin. No population numbers, no trends, I don't even know where the damn thing breeds, or what it eats. Do you?"

Management: "And we'll need it by the end of the month, okay?"

What can the poor guy do? Make some shaky assumptions about breeding, draw some wild inferences about habitat, grab some numbers from an old paper on the distantly related greater sculpin, stick them all into the mathematical model. Then hit "return" and hope for the best. Management gets a report with graphs and numbers they can quote to whomever asked them for "science advice." And Ken's smile gets a little more lopsided, because he's just done something a scientist shouldn't: come to a conclusion without actually testing his hypothesis. He's definitely not alone, because any time a scientist is asked to make a prediction and put numbers on it, he's doing exactly the same thing. Think climate change.

For me, the science boat I no longer want to sail in is like the noble four-masters of Burton's time: a few bones lying on the bottom, or a hulk hauled ashore and turned into tourist attraction. I'm glad of my scientific training because it provided me with a rational alternative to seeing the world as chaotic and unexplainable, but I don't envy those who do science now, especially when so much of it is modelling or thinly disguised technological fixes. We're going to see a lot more of these; the band-aids are already on the drawing board, like the vast sun-shading screens proposed to stop the bleaching of the Great Barrier Reef. The São Francisco Diversion project is just another one of these fixes, another example of the folly of trying to engineer ourselves out of a tight spot when the real solution is staring us in the face.

The trajectory of my own career in science was not so different from the path from water diversions to cisterns. I started out by passionately promoting the gene banking of threatened fish stocks, and it took me twenty years to realize that, while "the science" was right, the solution was as short-sighted as diverting a river. By the time I had followed the ups and downs of the São Francisco saga — painstakingly learned its long history, seen the battleground with my own eyes — it became clear to me that, while there were many lessons here, there would be no neat tying up of all the social, biological and historical threads, and my scientific training was no use at all. Maybe the diversion would get built, maybe it wouldn't;

In control: the nerve centre of Itaipu dam.

whether it did or not, or in what form, it was certain that others like it would keep popping up in other parts of the world. What really mattered about the São Francisco case, for planners and for concerned citizens, was the way the controversy brought out the essential injustice of the way water is shared in that part of the world, and how the tables could be, if not righted, at least turned partway using much simpler technologies.

The São Francisco experience also meant something more personal: it was the final confirmation that solutions based only on "the science" I was trained to revere are like the stiff belt of whisky at the end of a long day. They work, but not for long. As a way of understanding how the natural world works, science is the most gracious of muses: pure, unbiased, opening our eyes so that we can see clearly through to what is true. But you should never ask your muse to sleep with you, to try out some new positions. I don't think science can or should solve problems, no matter how many cabinet ministers "get it," certainly not those of the São Francisco River. Those ones could have been solved long ago using the tools Richard Burton already had. You don't need much science to build cisterns.

FAREWELLS

So ... "small is beautiful," is that the message? After dragging these people around with me, from the Philippines to Brejo in Brazil — Richard Burton, Barbara Tuchman, Miss Tojo — if that's the extent of their combined insight I might as well have read E.F. Schumacher in the 1970s and stayed home. Come to think of it, I did read his book, and I believed in his ideas; it just took me three decades to see for myself. Maybe Burton and Tuchman and even Miss Tojo helped me see a bit more clearly. What happened to them?

Before leaving South America, Sir Richard Burton went off on a sixteen hundred-kilometre overland trek across the Andes to Peru, down the coast through Chile and then around the Horn to Buenos Aires. He had no idea his wife had lobbied successfully for a new posting for him as consul in Damascus, nor that she had written a controversial preface to his forthcoming book on Brazil that sought to blunt his unpopular views on polygamy. In the end, the reviewers had a field day with the husband-and-wife team and their odd appearance of being at loggerheads, but *Explorations of the Highlands of the Brazil* was a bestseller. You can still buy it.

After the stint in Damascus, a time Burton called the happiest in his life, the inevitable recall crushed him. His last posting was as consul in Trieste, a slight appointment that lasted eleven years. He filled in his time with his labours on *The Kama Sutra of Vatsyayana* and the sixteen volumes of *Book of a Thousand Nights and a Night,* visits to his old haunts in India, and a short-lived enthusiasm for a disastrous gold mining venture in West Africa. He left most of the real work to his vice-consul as the British government officially threw up its hands and allowed him to have his way. He had become, finally, impossible to dismiss. In his infrequent visits to London, wrote one journalist, Burton "wakes up the Learned Societies, startles the Geographers, is the hero of banquets, and drops a new book in his wake."

Richard Burton died in Trieste in 1890, shortly after completing his new translation of *The Scented Garden of the Sheikh Nefzaoui,* the "crown of my life." He was sixty-nine and childless; he left an estate of 188 pounds sterling. Despite cash offers from publishers, his widow, Isabel, burned the only copies of *The Scented Garden;* whether for its extensive treatment of homosexuality or its quality as the last thing Burton wrote, biographers continue to argue. She was roundly hated for it, and publishers continued to hound her for the manuscript they believed she was hiding. But she was probably right to do it; this was the time of the imprisonment of Oscar Wilde, and Burton was no longer around to defend himself. She survived another six years, managing his literary estate.

Burton's extraordinary life makes me wonder: how many of us are really what we think or say we are? Burton the explorer lived his life disdaining the "little men" around him and their comfortable world. He didn't suffer fools. On arriving at Santos for his stint as British Consul, "one booby told me how comfortable were white clothes; I told him I had worn them before he was born." Spoken like a true iconoclast — but who did Burton really work for? His actual employers were the British East India Company and His Majesty's Government, as conservative a crew as one could hope to find, and Burton even spent some of that time as a secret agent. Straitjackets upon straitjackets, yet within their confines Burton managed to illuminate distant corners of all kinds. We should all admire him for that ability to make the most of his situation.

Many of us carry on our real business under a flag of convenience. Burton's flag was the company and the government; mine was biology. Burton travelled around pretending to be a consul; I pretended to be a biologist. For both of us, it was the unofficial notebook that filled up fastest. Maybe, at bottom, that's why he appealed to me in my twenties and he still does, prejudices and all, because he's one of that rare species of bird whose plumage is a disguise for something different, a hawk in the skin of a sparrow. Russian chemist Alexander Borodin, composing his seven symphonies; American insurance agent Charles Ives writing his *Concord Sonata;* transplanted banker T.S. Eliot writing The Waste Land or, more recently and more profitably, single mother J.K. Rowling sitting down at her kitchen table in Edinburgh to create *Harry Potter*. Good company for Richard Burton the British consul; all hawks, soaring above the rest.

And what of the São Francisco, the river that Burton the hawk circled, dove to investigate, then flew on from? Perhaps its identity, too, is still confused, despite its fervent embrace by post-colonial Brazilians as their own Velho Chico. It was those settlers that spread out along the São Francisco, and it was they (or their agents, the *bandeirantes)* who hunted down and enslaved the Indians living there. Those Indians, all those tribes that Burton characteristically describes, the Caetés, Tupinambá, Tapuya, Tupiaes, Amorpiras, Ubirajaras, they had a different name for the river. They called it *opará* — the ocean.

The Brazil Richard Burton had known and described so meticulously was already changing. Just three years after his death, Antônio Conselheiro and his band would settle in Canudos; in another four, they would all be dead too. There is a collection of photographs in Rio de Janeiro, taken by a photographer travelling with the final expeditionary force to Canudos in 1897; nowadays he would be called "embedded" in the troops. Flavio de Barros captured everything, including the slitting of the throat of The Counsellor's chief acolyte and the sawing off of The Counsellor's own head, which was sent to the University of Bahia for cranial study. (The learned phrenologists pronounced it, somewhat disappointedly, normal.)

There was a writer with the troops too, Euclides da Cunha, who documented the fall of Canudos as a war correspondent for the *Folha de São Paulo*. His resulting book, *Os Sertões* (Campaign in Canudos), has remained the classic description of the life of the poor *nordestino*. It's five hundred pages long; the edition I have is the twenty-first. Da Cunha himself went out with a bang too, killed in a duel with his wife's lover a decade later. In *The War of the End of the World,* the last image Vargas Llosa leaves the reader with is of the "thousands upon thousands" of vultures feeding on the remains and making a "a strange, indefinable, unfathomable sound, so loud it shook the air." Apocalyptic visions, all of them — reminders that, in the northeast of Brazil, things often end badly.

Barbara Tuchman went on to write one more book after *The March of Folly*, but for me and probably for many others, her second-last remains her best, because it explores specific historic incidents to illuminate a general, and very uncomfortable, truth about greed and self-interest. If Miss Tojo could read English, I would send her a copy.

In matters of fisheries and aquatic life, it's Miss Tojo we need to pay the most attention to now. I feel sure she is alive and well, still ordering her charges around corners and across intersections, still pursing her lips when they take a wrong turn. There's no reason to assume she's given up her sense of entitlement concerning fresh seafood either; as I write this she's probably looking up from an animated conversation with her cronies in a smoky Tokyo sushi restaurant, calling a peremptory *sumimasen!* to the waiter, ordering another plate of sea urchin roe, the new stuff just flown in from northern British Columbia, eaten straight out of the delicate globelike shell. Perhaps she'll call for a chilled Cabernet Sauvignon from grapes grown along the São Francisco River, to cleanse the palate.

I don't think Miss Tojo will ever stop, and that's what worries me. Yes, her country is the furthest ahead in making sure there's fish on the table, but Miss Tojo has friends all over the developed world, especially in Europe and North America. Every time I buy a bag of farmed Thai shrimp, fresh-frozen and conveniently machine-split along the back, I'm one of them. We like our seafood too, and our governments agree with Japan's that one of the best ways to get it is to manufacture the fish in hatcheries.

The rich red tuna Miss Tojo loves so much was once rightly considered the king of the sea, but where will it come from a decade from now? Already "the science" is being poured on the problem of tuna farming — not the crude feed-lots of Mexico and Australia and the Mediterranean but the egg-to-adult management of the entire life cycle. It's a brilliant technological fix for the very man-made problem of overfishing. Before long, Miss Tojo will be sitting on the deck of a cruise ship, happily eating excellent farmed tuna while the captain makes a discreet detour around the offshore cages.

What's it going to be: properly managed fisheries or hell-bent seeding of the oceans with man-made fingerlings? Multibillion-dollar retooling of entire watersheds or five-hundred-dollar cisterns? We can do the science on these questions and the thousands of others like them, but we shouldn't ask the science to tell us which path to choose. Only common sense and acceptance of our own humanity can do that.

THIS BOOK IS PART MEMOIR and part conservation science; it's the story of my response, as a scientist and as a person, to some unpleasant environmental realities. While the book contains some "hard facts," it's not a scholarly book or a work of journalism; to keep it readable, I chose not to lard it with citations.

Much of this book is based on personal notes on twenty years of fieldwork, in or on laboratories, boats, riverbanks, hotel rooms, fish farms, lecture halls, cars, airplanes, meeting rooms. In the beginning, I scribbled directly into notebooks; later, I taped my observations on a small digital recorder and transcribed them afterward. While some of the material and commentary on the São Francisco diversion is formally published, all of it, apart from a few news summaries, is in Portuguese. Web sites abound for the many issues raised in this book; I avoided taking "factual" information from them, relying instead on articles and reports I found on my own or that were kindly sent to me by colleagues in Brazil. There are, however, numerous published sources on waters, fisheries and the history and exploration of Brazil; the titles most useful to me are provided below.

Bestor, Theodore C. *Tsukiji: The Fish Market at the Center of the World.* University of California Press, Berkeley and Los Angeles, 2004. 411 pp.

Brodie, Fawn M. *The Devil Drives: A Life of Sir Richard Burton.* Norton, New York, 1967. 406 pp.

Burton, Isabel. *The Romance of Isabel Lady Burton.* eBooks@Adelaide, 2006. Originally published by Dodd Mead & Company in 1897.

Burton, Richard Francis. *Exploration of the Highlands of the Brazil, Volume II.* Elibron Classics (facsimile of the edition published by Tinsley Brothers, London, 1869). 478 pp.

Clarke, Robin, and Jannet King. *The Atlas of Water: Mapping the World's Most Critical Resource.* Earthscan, London, 2004. 127 pp.

Da Cunha, Euclides. *Os Sertões: Campanha de Canudos.* 21st edition. Ediouro, Rio de Janeiro, 2000. 526 pp.

De Villiers, Marq. *Water: The Fate of Our Most Precious Resource.* Stoddart, Toronto, 1999. 413 pp.

Godinho, Hugo and Alexandre. *Aguas, Peixes e Pescadores do São Francisco das Minas Gerais.* PUC Minas, Belo Horizonte, 2003. 458 pp.

Kerr, Alex. *Dogs and Demons: Tales From the Dark Side of Japan.* Hill and Wang, New York, 2001. 432 pp.

Lovell, Mary S. *A Rage To Live: A Biography of Richard and Isabel Burton.* W.W. Norton and Company, New York and London, 2000. 910 pp.

Melville, Herman. *Moby-Dick; or, The Whale.* Penguin Books, New York. 1013 pp.

Moorehead, Alan. *The White Nile.* Harper Perennial, New York, 2000. 448 pp.

McCully, Patrick. *Silenced Rivers: The Ecology and Politics of Large Dams.* Zed Books, London and New York, 2001. 359 pp.

Salgado, Sebastião. *Terra: Struggle of the Landless.* Phaidon Press, London, 1997. 143 pp.

Tuchman, Barbara. *The March of Folly: From Troy to Vietnam.* Ballantine Books, New York, 1984. 447 pp.

Vargas Llosa, Mario. *The War of the End of the World.* Translated by Helen R. Lane. Penguin Books, New York, 1997. 568 pp.

There are also, as one might expect, a great many Web sites related (and in some cases devoted to) the planned diversion of the São Francisco River. Most are in Portuguese. A small sampling includes:

Brazilian Semiarid Network (Articulação no Semi-Árido Brasileiro)
www.asabrasil.org.br
Information on the Brazilian semiarid zone, including the One Million Rural Cisterns Program.

Ecodebate
www.ecodebate.com.br
An environmental NGO that frequently posts information concerning the São Francisco.

Joaquim Nabuco Foundation (Fundação Joaquim Nabuco)
www.fundaj.gov.br
A federally funded NGO that promotes cultural and scientific activities aimed at the development of Brazilian society, especially in the north and northeast. Based in Recife.

Manuelzão Project (Projeto Manuelzão)
www.manuelzao.ufmg.br
Projeto Manuelzão is a Belo Horizonte-based NGO primarily concerned with the Rio das Velhas, a tributary to the São Francisco River. A respected voice with an academic focus.

Ministry of National Integration (Ministério da Integração Nacional)
www.mi.gov.br
The federal government ministry with oversight of the diversion project. Contains, among many other things, the project's Environmental Impact Report.

National Water Agency (Agência Nacional de Águas)
www.ana.gov.br

Contains a portfolio of projects intended to fight drought in the Brazilian semiarid region.

World Bank (Banco Mundial)
www.bancomundial.org.br
The section on The World Bank and the water sector in Brazil (O Banco Mundial e o setor de água no Brasil) contains discussion on transfer of water among hydrographic basins.

ANCIENT FOREST
FRIENDLY

69
trees were saved
for our forests

Preserving our environment

ECW Press Ltd. chose to use recycled paper
to print this book and saved these resources[1]:

energy	water	greenhouse gases	solid waste
63 million BTUs	112,983 L	4,698 kg	1,514 kg

Printed by **Webcom Inc.**
on Legacy Offset 100% post-consumer waste.

FSC

Recycled
Supporting responsible
use of forest resources

Cert no. SW-COC-002358
www.fsc.org
© 1996 Forest Stewardship Council

[1]Estimates were made using the Environmental Defense Paper Calculator.